Julia Walter-Herrmann, Corinne Büching (eds.)
FabLab

Cultural and Media Studies

Julia Walter-Herrmann, Corinne Büching (eds.)

FabLab

Of Machines, Makers and Inventors

[transcript]

This book arose at the University of Bremen in the working group »Digital Media in Education«, supported by the »VW-Foundation«, Hannover, GER.

Bibliographic information published by the Deutsche Nationalbibliothek
The Deutsche Nationalbibliothek lists this publication in the Deutsche Nationalbibliografie; detailed bibliographic data are available in the Internet at http://dnb.d-nb.de

The cover design by Kordula Röckenhaus (Bielefeld) is based on a photography of the ›Makerbot Thing-O-Matic‹ by Makerbot Industries (www.makerbot.com), source: www.flickr.com. This file is licensed under the Creative Commons Attribution 2.0 Generic license (http://creativecommons.org/licences/by/2.0/deed.en).
Composition and editorial by Julia Walter-Herrmann and Corinne Büching
Printed by Majuskel Medienproduktion GmbH, Wetzlar
ISBN 978-3-8376-2382-6

TABLE OF CONTENTS

INTRODUCTION

THE MOVEMENT

MATERIALITY AND VIRTUALITY

Maker Culture

Technology and Infrastructure

Community and Environment

Epilogue

NOTES ON FABLABS

JULIA WALTER-HERRMANN, CORINNE BÜCHING

Figure 1: Laser-cut and 3D printed objects by Oliver Niewiadomski at fab*digitalgardens in Bremen, Germany (Source: Photography by Justus Holzberger).

Koothrappali, PhD: You know, there is a way we can
get action figures to look exactly like us.
Wolowitz: Oh yeah? How's that?
Koothrappali, PhD: Two words: 3D printer! [...] they are an engineer's dream.
Anything you can design a 3D printer can make out of plastic [...]
Wolowitz: And we can make stuff we need for work with it:
prototypes of my CAD/CAM designs, specialized tools ...
Koothrappali, PhD: Not to mention 'Malibu Koothrappali'
in his totally bitchin' dream house.[1]
Big Bang Theory

"The digital culture's dynamics have led to a general acknowledgment of data production as the most important future option. However, the production of things seems to be outdated: Factories are not sexy!" (Boeing 2010, own translation) At the same time, there are developments and hints suggesting the digital future "lies outside the box, in making the box" (Gershenfeld 2005, p. 17). One will not be limited to making boxes, though. Since new technologies and machines enable people to easily produce chess pieces, jewelry, computers, batteries, teeth, yet action figures that look exactly like oneself (like proclaimed in the TV series Big Bang Theory) and all the other things one can imagine. The concept of turning ideas into things is probably as old as mankind. For a long time, one has been able to read and hear about enchanted lamps, mysterious stones and unknown cases that can make wishes come true and turn words into real objects. This fantasy has persisted over decades. In the 1980s, Star-Trek's spaceship Enterprise had a 'replicator' on board, a machine that could create any inanimate matter on demand.

In the present digital culture, digital data can transform into material objects and the formerly fictional idea of such a 'magic machine' has been turned into reality, namely by the further dissemination of small, digitally controlled production machines in FabLabs, so-called "labs for fabrication" (Gershenfeld 2005, p. 12), that are accessible for a broad public. These machines "are the pint-sized, low-cost descendants of factory-scale, mass manufacturing machines" (Lipson & Kurman 2010), for example 3D printers, laser cutters or CNC machines that produce objects on the basis of rapid prototyping, tooling and manufacturing (Chua et al. 2010, p. 18 et sqq.). Such production machines are able to print, cut or mill objects from data files without any human intervention.

Taking a look at the history and development of both fabrication devices and personal computers, one can imagine that digital fabrication devices will be accessible and used in everyday live in the near future. The first mainframe computers were huge, slow and expensive. To operate them, one needed to be an expert and nearly no one saw a general market for them. Computer pioneer Howard Aiken, a Harvard mathematician and creator of the Mark I calculator, even spoke of a demand of computers in total numbers of only five or six for all of the

1 | Taken from a dialogue between the characters of Rajesh Ramayan 'Raj' Koothrappali, PhD and Howard Joel Wolowitz from the CBS TV series Big Bang Theory, season 6, episode 14, first aired (USA), January 31st 2013.

USA (Ceruzzi 2003, p. 13). The story of digital manufacturing machines can be told likewise: Only twenty years ago, such hardware was huge, slow, expensive. To operate it, one needed to be an expert and nearly no one saw a general market for them. Back then, such machines were already in use in industrial manufacturing, but no one could ever imagine these machines getting established in private households or open accessible workshops. For a long time, people thought digital fabrication devices were only useful for the niche-economy of prototyping. Today such machines (some cost even less than $1000) can be found in every FabLab and even in some private households – and much more than just prototyping is done with them. At present, especially 3D printers, an essential part of every FabLab, increasingly get the media's and hence the public's attention. "A 3D printer is a computer peripheral[2] like any other, but instead of putting ink on paper, or data on a disk, it puts materials together to make objects" (Gershenfeld 1999, p. 65). The popularity of 3D printers can be explained as follows. On the one hand, 3D printers make it remarkably clear how an idea (or at least the virtual, digitally designed representation of an idea) can become a material object. On the other hand – particularly since there are affordable, easy-to-use, ready-made printers available in the market – this 'magic' now seems to be accessible for nearly everyone.

But FabLabs are neither chambers of magic nor mere accumulations of 3D printers and other fabrication devices. FabLabs are places where digital culture and material production merge and enter a new stage: There, one can find "collection[s] of commercially available machines and parts lined by software and processes [...] developed for making things" (Gershenfeld 2005, p. 12). These machines are based on digital technologies and operated with computers. Usually, a number of 'conventional' tools, like hammers, saws, and screwdrivers, materials, like plywood, glue, and cardboard, and small electronics, like micro controllers, LEDs, and little motors, are added to the collection of machines in these workshops. In these facilities, people can create material objects that can be beautiful or practical, complex or simple, 'intelligent' or not. FabLabs are open for interested individuals, such as artists, hobbyists and students, but also for entrepreneurs who want to "move more quickly from an idea or concept to a physical object or prototype, or [...] want to experiment with and enhance their practical knowledge of electronics, CADCAM[3], design, 21st century DIY" (Eychenne 2012, p. 5). The software used in FabLabs is usually available under Open Source (or comparable) licenses and therefore adaptable and developable (Delio 2004). Furthermore, a credo amongst "Fabbers" (Neef, Burmeister & Krempl 2005) advocates sharing the developed ideas among FabLabs and fabbers (Fab Charter 2012), mainly in the form of CAD files that are the prerequisites for the production of material objects. In doing so, a wide network of FabLabs around the globe, fabbers and files on various Internet platforms has already been established (Center for Bits and Atoms 2012). In this

2 | According to Eisenberg the commonly used expression 'peripheral' is not a well-chosen term for the promotion of such machines. He says, peripheral brings forth the idea of 3D printers and other manufacturing machines as being something unimportant, especially in comparison to 'the center of attraction' the computer itself (Eisenberg 2008, p. 62; explanatory note by the editors).

3 | CAD is an abbreviation for Computer Aided Design, whereas CAM is a common abbreviation for Computer Aided Manufacturing.

sense, FabLabs are globally connected, open workshops, where people can meet, collaborate, interact and exchange ideas, machines, tools, materials and software with the common purpose of making distinctive and digitally designed objects (from scratch) in an easy accessible and cheap way.

Neil Gershenfeld, physicist at the Massachusetts Institute of Technology's (MIT) Center for Bits and Atoms (CBA), USA, invented the concept of assembling modest production machines in small workshops for enabling everyone to make "almost anything" (Gershenfeld 2005, p. ix). The scientist installed the first FabLab in 2002[4] near his home university at the South End Technology Center in Boston, being supported by the National Science Foundation of the USA (Gershenfeld 2005, p. 25; Nunez 2010, p. 23). In 1998, Gershenfeld first offered a university course with the title How to Make (Almost) Anything, based on the use of professional production machines. "The workshops were designed for advanced Physical Sciences students in the throes of their research and promised to provide much needed experience on the kinds of high-tech fabrication tools" (Turner 2010, p. 29). When eventually, more than a hundred students signed up for the class, of which only a few had a background or at least any knowledge in 'cutting-edge' Physics and fabrication technologies, Gershenfeld started to wonder what all the architects and artists were doing in his class that had been planned for only ten students. The course instructor was even more surprised that the students' motivation to take the class was rather personal than scientific. The students wished to create "things they'd always wanted, but that didn't exist" (Gershenfeld 2005, p. 6), like missing or broken pieces of alarm clocks or 'artistic extravaganzas'. Surprisingly, all students managed to complete the course, dealing with the design, the use of computer-controlled machines and even the compulsory circuit building. They accomplished the course by spreading and exchanging knowledge within the huge and heterogeneous group. "The learning process was driven by the demand for, rather than the supply of, knowledge" (Gershenfeld 2005, p. 7), clarifies Gershenfeld. When the same scenario re-appeared year after year in his 'maker class', he realized the potential of a get-together of high-tech production machines with heterogeneous audiences and further developed the idea of establishing a permanent FabLab outside MIT, providing opportunities for tinkering, learning and creating for everyone. That was the moment FabLabs were born (Gershenfeld 2005, pp. 4-12)[5].

4| Different authors name various origins and commencing dates of FabLabs, mostly depending on the discourses and movements they relate themselves to, such as hacker- or Open Hardware movements. All authors of this book refer to the labs that arouse in the outreach of MIT's CBA.

5| Meanwhile, the course How To Make (almost) Anything is available online at the CBA's website, last viewed 15 January 2013 <http://fab.cba.mit.edu/classes/MIT/863.08/>, so that everybody who is interested can take Gershenfeld's class independent of being an MIT student. The web seminar is also an essential part of many FabLabs around the world, where the lessons are streamed via Internet on a weekly basis. The How To Make (Almost) Anything course offers instructions for students and interested people about digital fabrication and the use of high-tech manufacturing tools. The seminar is part of the Fab Academy, an online outreach program of the CBA that can be visited here, last viewed 15 January 2013 < http://www.fabacademy.org/>.

From the outset, Gershenfeld's fundamental idea was not only "to make (almost) anything" (Gershenfeld 2005, p. ix), but to make fabrication technologies accessible for 'almost anybody' and hence empower people to "start their own technological futures" (Gershenfeld 2005, p. 17). He states that we "had a digital revolution, but we don't need to keep having it. Personal fabrication will bring the programmability of the digital worlds we've intended to the physical world we inhabit" (Gershenfeld 2005, p. 17). The scientist compares the development of FabLabs with the rise of the Web 2.0, when tools and applications for composing, editing and sharing digital content online became increasingly available for everyone, turning users into prosumers. In FabLabs, the possibilities of digital fabrication become further accessible and prosumers can compose, edit and share (designs for) material artifacts (Gershenfeld 2006). These potentials are exponentiated by the idea of a FabLab as an early version of a "Personal Fabricator" (Neef, Burmeister & Krempl 2005, p. 20; Gershenfeld 1999, p. 64 et sqq.), a digital production machine at home. Such an expansion could have an enormous impact on the value of things, communal life, or even whole economies. By all means, Gershenfeld understands FabLabs and related technical progresses rather as a "concept for development" (Boeing 2010; own translation) than simply as high-tech production laboratories. FabLabs shall stand for a concept of reducing the uneven distribution between the few producers and the many consumers or at least herald a future that links itself to a pre-industrialized past: "Such a future really represents a return to our industrial roots, before art was separated from artisans, when production was done for individuals rather than the masses" (Gershenfeld 2005, p. 8). The idea of an individualization and democratization of (the means of) production caused the establishment of further FabLabs in India in 2002 and in Ghana in 2004 (Delio 2004), where people should be supported in producing things of personal need and desire and therefore reduce economic dependencies and develop a 'subsistent freedom'. A 'doing good factor' doubtlessly is an essential part of the approximately 120 FabLabs on five continents (Center for Bits and Atoms 2012).

Right from the start, all FabLabs have been operated based on the same basic principles "to empower, to educate, and to create 'almost anything'" (Nunez 2010, p. 24; his emphasis). This belief was already put on record by the CBA in the Fab Charter, sort of the FabLabs' 'constitution'. The Fab Charter furthermore sheds light on additional FabLab relevant aspects, such as open access to labs and machines for everyone, responsibility for own actions, machines and environment, free knowledge dissemination, the protection of intellectual property rights and the sustainability of FabLab activities (Fab Charter 2012). Since the establishment of the first FabLab, field practitioners and laboratory researchers gather regularly for various meetings. The International Fab Lab Forum and Symposium on Digital Fabrication takes place at different FabLabs around the globe each year (Center for Bits and Atoms 2012). These conferences are strongly supported by the International Fab Lab Association that was officially established in 2011. The Fab Lab Association is an association of around 200 active and dedicated FabLab members that aim at serving the FabLab community by sharing their experience working with digital fabrication and organize the widespread FabLabs and individuals (International Fab Lab Association 2012).

At present various authors enthusiastically declare the world to be in a phase of transition. "The New Industrial Revolution" (Anderson 2012) and the end of mass production are proclaimed likewise. Such predictions mainly draw on the increasing availability of new ways and machines for production, similar to those in FabLabs. MIT's Technology Review even set up a blog section about the topic, where the "Next Wave of Manufacturing" (Technology Review 2013) and a "manufacturing renaissance" (Technology Review 2013) were announced, thus the blog critically advices companies to "invent the manufacturing technology of tomorrow" (Technology Review 2013). However, the impacts of FabLabs spread into many different social fields, not only into the techno-economic sphere. In times of a digital culture and increasing individualization within changing societies, FabLabs are important places for corporate learning, working and playing with advanced technologies. Being a global movement and part of a rising maker culture, FabLabs are central for an understanding of the present (and future) world. The democratization of production comes along with a 'democratization of innovation' by various potential actors. That means that, in FabLabs, everybody can invent, create and modify things and everybody can become an artist. With relatively low constraints, people can design objects that are not only unique, but meet high design standards, too. Such an approach transforms the fields of arts and crafts, as FabLabs further promote an understanding of modern crafting, making, or DIY as a response to mass culture. Despite the potential of democratization of innovation through FabLabs, a frequently referenced concern focuses on the diversity of potential actors[6]. It should be taken into account that not only academic urban males in their late twenties participate in the FabLab culture. FabLabs may create initiatives to invite economically and socially disadvantaged people to FabLabs, e.g., by organizing special workshops for marginalized people. Another relevant aspect of FabLabs stresses their potential for learning that was already put down in the Fab Charter. In order to establish a creative culture of making instead of copying, FabLab-based activities may also be included in school curricula for problem-based learning, creative hands-on activities and developing skills for documenting and communicating ideas and problems efficiently.

But even if the praises and promises for FabLabs are high at the moment, new techniques and technologies never appear without contempt, criticism and fear. Aspects such as copyright – which have mainly affected music and filmmakers until now – will affect the manufacturing sector, too. In a world where one can remotely print the same objects virtually everywhere, this will not only develop international collaboration, but also challenge limitations of national legislation. The advantage that one can print his/her own spare parts to replace the broken original parts will bring about issues such as security, liability and warranty. The cases of printable weapons and digitally manufactured food incite discussions about the power of technology and user ethics. Meanwhile, many FabLab practitioners and activists are concerned with establishing business models for their FabLabs and improving the organizational structures supporting a global community.

6 | Various speakers at the conference FabLearn – Transformative Learning Technologies Lab, 2012, in Palo Alto, Cal., USA, raised these concerns. For more information see the website, last viewed 25 December 2012 <http://tltl.stanford.edu/fablearn2012>.

Ten years after the first FabLab opened, it is time to look back on a decade of FabLab activities and enrich the number of written academic research about it. The sustaining of academic standards and scientific perspectives on the phenomenon of FabLabs is an essential feature of the several articles of the following book; the pictures and illustrations pay tribute to the design approach of FabLabs. Far beyond any revolutionary rhetoric, this collection seeks to scientifically analyze FabLabs and its entailed cultural and social changes with its foundations and impacts. It is not meant to explain any potential new world order(s) but to precisely analyze existing FabLab-related phenomena and to conclude their significance for present and future societies. Not only does it work as a compendium about various aspects FabLabs deal with – the delectable ones as well as the critical ones – but it also tries to show how a range of topics can be negotiated in/with FabLabs and what their theoretical foundations are. In doing so, 'FabLab – Of Machines, Makers and Inventors' is no technological report, but rather an analysis of the present times, discussing various approaches to and social and cultural impacts of FabLabs. In this composition, fabbers, scientists and designers reflect their perspective, experience and knowledge, discuss relevant theoretical and empirical questions and problems and introduce practical methods and outcomes of digital fabrication. In this volume, divers FabLab experts, either practitioners or field researchers, give insides to FabLab related issues, knowledge and organizational structures. Since FabLabs are continuously developing and the range of relevant and related aspects seems to be endless, the list of authors and topics must remain incomplete.

This volume is divided into five sections (and an epilogue) according to the varying foci in order to systemize the various FabLab-relevant issues. They are called 'The Movement', 'Materiality and Virtuality', 'Maker Culture', 'Technology and Infrastructure', 'Community and Environment'. Each section begins with a 'Notes on ...' chapter, in which experts of a specific issue report their very personal experiences and subjective view on a certain topic. The further chapters in each case discuss one aspect of an issue under a specific perspective, either on a theoretical or empirical base, or based on experiences in the field.

The first section is called 'The Movement' and deals with aspects that concern all FabLabs or that are mostly related to the global association or sum of FabLabs. It describes how different FabLabs work together, how they are linked, which aspects are relevant for the establishment of a FabLab or how a FabLab can focus on a certain aspect, like gender or empowerment for children.

The section begins with 'Notes on The Movement' by Karsten Joost. Karsten Joost from Bremen, Germany, was born into a family of craftsmen, he is a toolmaker by profession, and defines himself as an artist and networker, too. In his notes he describes his fascination with FabLabs and its origin, while he also expresses an authentic concern about the difficulties he sees in building a FabLab from scratch.

Julia Walter-Herrmann, a researcher in the working group Digital Media in Education (dimeb) at the University of Bremen, Germany, talks about 'FabLabs – A Global Social Movement?' She states that hardly any debate about FabLabs understands FabLabs only as a collection of machines that is resembled in small-scale workshops; most of the texts and lectures about FabLabs also grasp the global alliance of FabLabs under the keyword 'FabLab'. Her text raises the question whether FabLabs can be called a global social movement. Therefore, Walter-Herrmann's text briefly introduces the characteristics of a movement and then follows up with a theoretical and empirical analysis of the worldwide network of FabLabs.

Lambert Grosskopf, who is a legal academic at the University of Bremen, Germany and an Attorney at Law, a Certified Lawyer for Information Technology Law (IT-Law) and a Certified Lawyer for Copyright and Media Law, writes about copyright issues concerning 3D printing, using the example of Germany. In his article 'Homo Fabber and the Law', he claims that the use of 3D printers is engendering an emergent conflict that makes current conflicts of interest among Internet users, copyright holders and exploiters of intellectual property over the issue of file sharing seem like a playground tussle. Soon it will not be the object itself that is swapped over the Internet, but the CAD data file for its production. Grosskopf elucidates the problem set that evolves when a consumer transforms into a prosumer who produces products that are originally protected by copyright, patents, utility models and registered designs autonomously in his or her basement hobby room.

Tanja Carstensen from Hamburg University of Technology, Germany, has been researching gender and various digital technologies for several years. In her contribution 'Gendered FabLabs?' she argues that traditionally, technology has been linked with power and masculinity. In her work she examines which roles gender and the strong connection between technology and masculinity play in FabLabs as high-tech spaces. Following the idea that every new technology opens new possibilities to negotiate gender roles and new gender relations, she analyzes the gendering of FabLabs according to the following categories: access, users, technology, products, education, community and empowerment. In conclusion, she discusses the opportunities for shifts in gender and technological relations.

Irene Posch works for the HappyLab in Vienna, Austria, and previously worked on building the FabLab at the Ars Electronica Center Linz. In her chapter 'Fabricating Environments for Children' she writes about the specific requirements for FabLabs when working with children. She declares that, when it comes to children, an open lab and self-directed access based on peer-to-peer learning are often not enough, as most children do not have the possibility to learn about and with these technologies at any other place. Her article gives an insight into the practical work with children at the local FabLab in Vienna, Austria. The people from the Vienna HappyLab set up introduction workshops as well as an afternoon program allowing children to come to the lab and independently work on their individual projects, while getting professional help where wanted and needed. Their goal is to provide an understanding of the lab's possibilities and allow for personal creation in order to relay ideas that facilitate active and informed use. Posch then reports on her experiences, focusing on the interest and understanding children developed towards the introduced digital fabrication technologies.

The subsequent section is about 'Materiality and Virtuality'. The section's focus lies on the relationship between ideas and their virtual representations and real, materialized, and graspable objects. It presents mathematics and coding, thus the foundation of software and computers, as forms of complex or syntactical arrangements of such immaterial ideas. It asks whether the relationship between materiality and virtuality has changed, now that one can print, mill or cut any (three-dimensional) idea. It furthermore asks whether such an alteration could be crucial for learning. The section ends with an analysis of material objects that are produced in FabLabs.

'Notes on Materiality and Virtuality' by science fiction author Bruce Sterling open this section. Sterling is a visionary of the digital sphere, having published various well-acknowledged novels and texts about a techno-culture and helped to define the cyberpunk genre. Furthermore, he writes a blog hosted by Wired magazine. For his contribution, he referenced the neologism 'spime', which he invented in his book Shaping Things. Spimes are objects that can track their history and interact with the world. In his notes, Sterling explains what this has got to do with FabLabs, while also speaking of sustainability and aspects of a 'healthy ecology'.

Frieder Nake is a professor emeritus from the University of the Arts Bremen and the University of Bremen, Germany. Nake is one of the founding fathers of (digital) computer art. He is a mathematician and a computer scientist, and his dedication and unique approach to scientific problems gave him the sobriquet 'poet of sciences'. He won several awards – not only for his artistic work, but also for his extraordinary teaching. In his text 'Considering Algorithmics and Aesthetics', he explains that algorithms are statements of generality. In contrast, works of art are statements of particularity. The two kinds of statements differ in some more respects, but in the digital domain they must come closer and, in fact, unite. Here, algorithmics and aesthetics should support each other. Nake's essay recalls the historic moment in 1965 when the generative art movement started. It characterizes the ontological status of art and technology and furthermore hints at the dialectical unity of virtuality and actuality. Examples – amongst it an example by famous computer artist Casey Reas – demonstrate how the algorithmic and the aesthetic tendencies of today's world have begun to meet.

The subsequent contribution comes from Heidi Schelhowe, who is a university professor for Digital Media in Education (dimeb) and the head of a working group of the same name. Schelhowe is an internationally renowned researcher and the Vice Rector for Teaching at the University of Bremen. Her research focuses on educational applications in Computer Science and Media Informatics, as well as on digital media and media education within the context of pedagogical didactics. In her contribution 'Digital Realities, Physical Action and Deep Learning – FabLabs as Educational Environments?', she discusses the educational potential that arises from the special alliance of virtuality and materiality in FabLabs. She presents a brief history of learning materials and arrangements and its role for society. Referring to this history, she explores the very specific and original benefit that FabLabs can offer in comparison to other educational environments from a general and theoretical point of view.

The section 'Materiality and Virtuality' ends with an article by Corinne Büching about 'A Universe of Objects'. Büching is a sociologist in the field of science and technology. She is a research assistant in the computer science's working group dimeb at the University of Bremen, Germany. In her text, she first introduces a concept about the essence of (digitally produced) objects, furthermore demonstrating how these objects can be empirically analyzed. She concludes her text with the analysis of objects that evolved from a workshop she gave at the FabLab St. Pauli.

The third section titled 'Maker Culture' centers on the production of things in workshops and in an artisan and hand-made way in contrast to the common mass production in factories. It further explores how the formerly separated spheres of professional design and amateurish craft are currently negotiated with the introduction of digital production machines into both spheres. Catchwords that are relevant for this section are, for example, empowerment, DIY, crafting, tinkering and design. The section discusses aspects of professionalism and design in (digital) manufacturing, but also the history of making and producing things with the support of computers.

The section starts with 'Notes on Maker Culture' by Eva-Sophie Katterfeldt, Anja Zeising and Michael Lund. They are all researchers from the working group dimeb and have various backgrounds in Digital Media, Computer Science, Art and Cultural Studies. From these various standpoints, they regard the topic of maker culture, briefly explain the origin of the term and relate it to other aspects of maker culture such as interaction design and everyday crafting.

The second text was also contributed by a dimeb-member, Bernard Robben, a senior researcher at the University of Bremen. He has published various articles on media theory, the computer as a medium, tangible embedded and embodied interaction and the design of 'be-greifbar' (tangible and graspable) media. In his article, he presents 'The History of Production with Computers' in order to carve out their potential for digital fabrication. The paper brings into focus the close relationship between the evolution of production machinery and of plans, models, drawings, and diagrams. It describes the interweavement of physical machinery and the corresponding virtual one within the context of processes of technological and social change.

The chapter 'Maker Culture, Digital Tools and Exploration Support for FabLabs' by dimeb-researcher Eva-Sophie Katterfeldt introduces the so-called 'maker culture' in which FabLabs are situated. She explains what this maker movement distinguishes from previous DIY cultures is the involvement of digital media in the creation process – as web platforms for information retrieval, communication and sharing, as tools for digital crafting, as well as target artifacts of creation processes. Open sharing, creativity, learning and participation are values of the maker culture. FabLabs offer this whole range of involvement with digital media, inviting everyone to participate independent of skill. Katterfeldt argues that in order to make these opportunities better examinable for every maker the design of digital crafting tools should be rethought.

Jens Dyvik, a cabinetmaker and designer from Norway, who is currently on a FabLab world tour, wrote the article 'Thoughts from the Road of a FabLab Nomad' when he was visiting the FabLab in Indonesia. Dyvik's goal is to research personal manufacturing and Open Source design around the world. The findings of his research will be communicated through a documentary called 'Making Living Sharing'. In his text, based on his experiences, he asks how designers can support people in creating their own products. Moreover, he discusses whether designers can still make a living if they shared their designs with the world. Dyvik is working towards answers to his research goals by visiting FabLabs around the world and by learning how they facilitate knowledge-sharing and personal production.

The fourth section of this book, 'Technology and Infrastructure', is dedicated to the technology-relevant aspects of FabLabs. On the one hand, it explains individual machines, such as 3D printers, laser cutters and CNC mills; on the other hand, it also describes how the produced artifacts can again be enriched with several technologies like microcontrollers. Furthermore, it relates technological characteristics to qualities of a technological infrastructure. Aspects of digitization already find their expression in the philosophy of Open Source for the production and development of software and even hardware, but the section also raises critical concerns when it comes to turns of organizing or financing these technological infrastructures.

The section starts with 'Notes on Technology and Infrastructure' from Bre Pettis, co-founder and CEO of MakerBot®, a Brooklyn-based company for 3D printers, and of Thingiverse, an internet platform where people can share their 3D printing files. MakerBot's new, low-cost desktop 3D printer, the Replicator 2, made it into media coverage all over the world, and Pettis was even represented on the cover of Wired magazine. In his text, he introduces his company and the machines it sells, he furthermore envisions a future of 3D printing.

The next article, 'Machines for Personal Fabrication', comes from René Bohne. Bohne is a research assistant at the Media Computing Group at the Rheinisch Technische Hochschule Aachen, Germany. He is also the local FabLab manager at his university. Bohne is highly interested in personal fabrication and personal design as well as wearable computing and smart fashion. For this book, he introduces tools for digital fabrication used in FabLabs today. In addition, his paper describes tools for personal fabrication at home. Explaining, for example, the technical differences between additional and subtractive fabrication, this contribution works at the same time as an introductory text for FabLab novices and technically interested people.

The following chapter, 'Digital Fabrication in Educational Contexts – Ideas For a Constructionist Workshop Setting' by Nadine Dittert and Dennis Krannich, presents practical efforts combining FabLabs with tangible technology. Dittert and Krannich are both research assistants in the working group dimeb at the University of Bremen. They are computer scientists who are not only interested in setting up a FabLab at their local university, but also in developing learning and teaching scenarios for FabLabs. In their article, they discuss their experiences with 'teaching FabLabs'. As an example, they show the 'Fab-tast-O-matic' – a chain reaction machine (inspired by Rube Goldberg's machines) – that was built in a student's Bachelor project. The machine consists of four different layers that are equipped with Arduino micro controllers and electronic components, 3D printed and laser-cut parts, and other tinkering materials. Their assumption is that digital fabrication enables people to explore how to represent a functional description of a system by physical shapes, and ask in return to which extent a functional description of a physical system can be abstracted.

Peter Troxler contributed the last text in the section 'Technology and Infrastructure'. Troxler is often referred to be the mastermind of the European FabLab movement; he furthermore is a Research Professor at Rotterdam University of Applied Sciences and a freelance researcher at the joint of business administration, society and technology. In his article 'Making the Third Industrial Revolution – The Struggle for Polycentric Structures and a New Peer-Production Commons in the FabLab Community' Troxler addresses how economic models and cycles of innovation change, based on new ways of producing things, such as peer production or Open Design, and organizational structures and models, such as the Open Source movement, or FabLabs. But the chapter also aims to provide more than just a definition of the problem of the struggle for more polycentric structures. It also addresses how to build effective forms of collective action and self-organization for the FabLab community and how to break free from traditional systems and creatively design new systems that tap into the capabilities of that community.

Section five is called 'Community and Environment'. This part deals with examples of FabLabs that are embedded in a (local) community, aiming at improving the status quo. It shows how individuals and groups can benefit from FabLabs. The section illustrates several best practices of FabLabs around the world and how they empower, educate, and create 'almost anything'. It displays the impact of FabLabs for the field of learning, neighborhood development, health, and cooperation.

Bart Bakker from the Netherlands opens the section with 'Notes on Community and Environment'. Bakker is deeply involved with Protospace, the main FabLab in Utrecht, Netherlands, and with the Dutch FabLab Society. In addition, Bart Bakker set up a mini FabLab in his garage in Utrecht. He constantly invites artists, engineers, and the interested public to visit his 18 m^2 space, where his FabLab for cutting and printing objects is set up between his model railroad and his car. In his notes, he talks about his experiences with FabLab machines and describes his experiences with engaging in a project for Open Source laser cutters.

The first chapter of the section is entitled 'Digital Fabrication and 'Making' in Education: The Democratization of Invention'. It is a contribution from Paulo Blikstein, who is an Assistant Professor at Stanford University, USA. He is affiliated to the Stanford University's Graduate School of Education and (by courtesy) Computer Science Department. He designs and researches expressive technologies for learning, especially for underprivileged people. In his text he classifies learning experiences with FabLabs in the tradition of 'unconventional pedagogy' as so evolved by John Dewey, Seymour Papert, and Paulo Freire. He argues that digital fabrication and making could be a new and major chapter in the process of bringing powerful ideas, literacies, and expressive tools to children. Using examples of implemented FabLab learning scenarios Blikstein highlights five important design principles.

Axel Sylvester and Tanja Döring contribute the next chapter to the section 'Community and Environment'. Sylvester and Döring are involved in the FabLab in their hometown Hamburg, specifically a borough called St. Pauli, in Germany. Sylvester is an IT consultant and independent researcher, he is also engaged in the global Fab Lab Association. Döring is a computer scientist, who works as a research assistant at the University of Bremen at the Center for Computing and Communication Technologies. Their article 'Urban Development with FabLabs' describes the potential they see for the role of FabLabs in a larger network of innovation, and applies urban development patterns to FabLabs on the basis of experiences from the FabLab Fabulous St. Pauli in Hamburg.

In the third chapter, the manager of the FabLab at Vigyan Ashram in Pabal, India, Yogesh Ramesh Kulkarni, explains the particular challenges and specific opportunities a FabLab in rural India has (to face). In his text 'Small Ideas, Big Opportunities – FabLab at Vigyan Ashram Pabal, India', he introduces the history of this FabLab, its educational approaches, how the FabLab blends with 'traditional' manufacturing tools, as well as some successful stories, which were 'shaped' through the FabLab at Vigyan Ashram. He furthermore argues that one distinctive feature of FabLabs in rural India is that it can help to solve 'grass-root problems' of local villagers. In doing so, students cannot only help to find unique solutions to specific local problems, but they can also learn something important for their lives and get empowered through a practical hands-on learning experience.

Alex Schaub wrote the subsequent contribution about 'Affordable Medical Prostheses Created in FabLabs'. Schaub is the FabLab manager of Waag Society's FabLab in Amsterdam in the Netherlands. Furthermore, he is the head of the $50 Leg Prosthesis Project in Yogyakarta in Indonesia. In his chapter, Schaub describes that developing a $50 below-knee prosthesis is a challenge. Moreover, he questions whether it is even possible, considering that a below-knee prosthesis costs $4,000 in the Western world. The chapter explains how Waag Society's FabLab Amsterdam and the FabLab Yogyakarta in Indonesia, run by the media and art laboratory House of Natural Fiber, are trying to turn such a challenge into reality.

The epilogue 'FabLabs: Thoughts and Remembrances' is written by Sherry Lassiter. She is the Director of the international Fab Foundation and Program Manager for the Center of Bits and Atoms at the MIT. Lassiter supports and coordinates the worldwide FabLab network. Her text leads us back to the place where FabLabs were 'born' – the MIT. Ten years after the first FabLab opened, Lassiter reviews the last ten years of FabLab activities and introduces various projects in the outreach of CBA from around the world.

The idea for this compilation arose during a research project called Subject Formations and Digital Culture that was funded by the VW-Foundation in Hannover, Germany. At one stage of the project, we empirically researched learning in interaction with digital artifacts and therefore conducted workshops with young adults in FabLabs. Besides conducting our empirical research, we also developed a growing fascination for FabLabs that was shared by our working group Digital Media in Education (dimeb) at the Center for Computing and Communication Technologies at the University of Bremen, Germany. While those co-workers who were more interested in technology started to build 3D printers and tinkered with them, we, as social scientists, were more interested in FabLabs as a social phenomenon, asking ourselves: What possibilities for economy, learning, and crafting do FabLabs offer? Has the relationship between virtual and material objects changed through FabLabs, and what is the motivation for fabbers to engage in the labs? How will the future of FabLabs look like? These were all questions that kept us up all night. While discussing these questions with the head of dimeb, Heidi Schelhowe, we recognized that there was no person, book or article that could answer all our inquiries. In addition, we realized that there were hardly any scientific texts addressing our questions – except of course the fundamental books of Neil Gershenfeld, which had been published years before. For this reason, without further ado, we decided to gather the existing FabLab knowledge and edit it for this collected edition. However, as is usual with any project, no one can master it alone, which is why we would like to thank all those helpful hands and clever minds that supported us.

First, we would like to thank the VW-Foundation for their confidence in our idea and their generous financial support, without which this project would not have been possible. Furthermore, we are very grateful to Heidi Schelhowe, who always encouraged our intentions and enriched the book with her precious advice, critique and experience. We are also in debt for gratitude to the people of dimeb, who were not only very patient in introducing us to and teaching us the FabLab technologies, but also made lots of helpful suggestions for the composition of the book. Our last thanks we would like to give to our student assistants Lea and Camila. A special thank goes to Michael Lund, Dennis Herrmann and Michael Dorschner – they know why.

We now hope that the articles in this collection offer valuable information and enrich the debate about FabLabs in a constructive way.

References

Anderson, C 2012, *Makers: The New Industrial Revolution*, Crown Business, New York.

Boeing, N 2010, *The Future is Fab*, media release, 3 March, Heise Online, viewed 10 January 2013, <http://www.heise.de/tr/blog/artikel/The-future-is-fab- 942575.html>.

Center for Bits and Atoms 2012, viewed 11 January 2013, <http://cba.mit.edu/>.

Ceruzzi, P 2003, *The History of Modern Computing*, MIT Press, Cambridge, MA.

Chua, CK, Leong, KF & Lim, CS 2010, *Rapid Prototyping: Principles and Applications*, World Scientific Publishing, Singapore.

Delio, M 2004, *Ghana Gets a Fab Lab*, media release, 9 October, Wired, viewed 10 January 2013, <http://www.wired.com/science/discoveries/news/2004/09/64864?currentPage=all>.

Eisenberg, M & Buechley, L 2008, 'Pervasive Fabrication: Making Construction Ubiquitous in Education', *Journal of Software*, vol. 3, no. 4, pp. 62-68.

Eychenne, F 2012, *Fab Labs Overview*, viewed 10 January 2013, <http://de.slideshare.net/slidesharefing/fab-labs-overview>.

Fab Charter 2012, viewed 11 January 2013, <http://fab.cba.mit.edu/about/charter/>.

Gershenfeld, N 1999, *When things start to think*, Henry Holt and Co, New York.

Gershenfeld, N 2005, *Fab. The coming revolution on your desktop – from personal computers to personal fabrication*, Basic Books, New York.

Gershenfeld, N 2006, *Fab Labs*, TEDTalk, viewed 2 January 2013, <http://blog.ted.com/2009/08/11/fab_labs_neil_g/>.

International Fab Lab Association 2012, viewed 11 January 2013, <http://fablabinternational.org/>.

Lipson, H & Kurman, M 2010, Factory@Home. *The emerging economy of personal manufacturing. Overview and Recommendations, US Office of Science and Technology Policy*, viewed 10 January 2013, < http://web.mae.cornell.edu/lipson/FactoryAtHome.pdf>.

Neef, A, Burmeister, K & Krempl, S 2005, *Vom Personal Computer zum Personal Fabricator – Points of Fab, Fabbing Society, Homo Fabber*, Murmann, Hamburg.

Nunez, JG 2010, 'Prefab the FabLab: Rethinking the Habitability of a Fabrication Lab by Including Fixture-based Components', Master's thesis, Massachusetts Institute of Technology, Cambridge, MA, viewed 10 January 2012, <http://dspace.mit.edu/handle/1721.1/59201>.

Technology Review 2013, viewed 11 January 2013, <http://www.technologyreview.com/>.

Turner, R 2010, 'Open Source as a Tool for Communal Technology Development: Using Appropriate Technology Criteria to Determine the Impact of Open Source Technologies on Communities as Delivered Through the Massachusetts Institute of Technology Fab Lab Projects', Master's thesis, Wits School of Arts, University of the Witwatersrand, Johannesburg.

THE MOVEMENT

NOTES ON THE MOVEMENT

KARSTEN JOOST

Figure 1: Impressions from various FabLab events in Bremen, Germany (Source: Photography by Justus Holzberger).

My name is Karsten Joost, and I was born in Verden, Germany in 1968. I am a toolmaker, artist and networker. I heard about FabLabs for the first time from friends in Groningen (Netherlands) in 2008, as they had plans to develop a FabLab in their city. In 2010, they opened the doors to their lab. To be honest, I needed quite a long time to develop an understanding of what FabLabs are about and why a worldwide community would keep such an interest in an assemblage of computers, several machines and soldering guns, whilst talking about the next industrial revolution and personal fabrication. Anyway, once I got involved with it, I started to understand its fascination. Most of my personal motivation to support the FabLab idea is derived from my biography. I grew up in a family of craftsmen: my grandfather was a blacksmith and my father repaired trucks. The first profession I learnt was toolmaker. Much later on, I studied Fine Arts, and I always felt that these fields are too disconnected from each other. However, FabLabs provide a space where both can exist side by side and this joint working makes sense to me. An idealistic goal for me is to integrate art more closely to what one might call the maker or fabbing culture. From my point of view, artists are researchers who have much more in common with, for instance, engineers and scientists than the clichés tell us. Art is a language we should not miss. It is a language that brings people together. Hence I decided to set up a FabLab in Bremen, Germany, as the city has been my hometown since 1986, and I still find life here quite enjoyable. With about 550.000 inhabitants, the city is convenient to get around and the people's open-minded mentality makes it easy to find partners to develop a FabLab.

At first, I understand FabLabs as multidisciplinary learning environments. Behind these disciplines are institutions that I like to cooperate with when setting up and running the Bremen FabLab. I would like to develop a learning atmosphere in which institutions from science, business, education and culture could participate in an active way. For me, FabLabs seem to bring about a curious environment: The labs enable people to build things on their own, just like a craftsman in the Middle Ages, when mass production was a faraway thought. But at the same time people can act like a modern industrialist, or even more freely than him or her, because one can send a self-designed product as bits and bytes in any direction around the world to the people one likes. This makes me wonder: Do FabLabs mark the beginning of a new era? Besides finding machines that are able to re-produce things and themselves, I also found engineers and craftsmen working side by side with designers and artists (an aspect of FabLabs which touches me deeply) during my visit in The Netherlands. They inspire each other and share knowledge in an open-minded atmosphere. These high tech laboratories seemed to me like an interface for extraordinary collaboration. Beside a digital revolution, I felt confronted with a kind of Cultural Revolution, which keeps an enormous potential for innovations based on new forms of collaboration. Last but not least, FabLabs encourage the development of new forms of education.

An African saying reads, 'a whole village is required to provide a good education to a child', and I think for today's youth it is important to gain a feeling for the society in which we all live. The more transparent a society is to them, the more they can develop an idea about their own future within it. I believe this is a basic motivation for any young person to acquire knowledge. The FabLab community supports the sharing of inspiration and expertise. There is something very peaceful about this concept, especially for the global political context. Everybody is a specialist in something, who should be asked about his/her personal expertise and ideas, and should receive acknowledgement for this. Knowledge transfer is a very interesting idea for me. I would like to spend much more time thinking about it.

However, I do not only have positive memories about FabLab activities: I had a rather tough time during the Fab6 Conference in Amsterdam. I had built my tent close to the center of the event. It had only rained for three days in August but these were the three days when the event took place. The whole campground was muddy (I guess it was just like Max Yasgur's field during the Woodstock-festival). Never before had I felt that wet and smelly for a whole week! Still, at that time I was already convinced that FabLabs in general and the idea of building a FabLab in Bremen are worth camping in the rain. I think the most important thing is to find out what exactly motivates oneself to build a FabLab. The diversity of ideas in the FabLab community is extremely high. There are many different ways to interpret its meaning (which makes the community so rich). So I advice anybody who also likes to build a FabLab: Take your time to figure out what motivates you about FabLabs and check from time to time if you are on the right track. Furthermore, don't act in isolation. FabLab is a worldwide community, so look for partners.

My future vision for FabLabs is connected to my idea of the development of tomorrow's working environments. Many more jobs might be organized in forms of projects. There might be a goal and a deadline by when a project should be completed and people might decide for themselves how to reach this goal. Such a development provides people with more liberties but also with more responsibility. Good jobs might be given to those people who have got enough expertise and who do not hesitate to start collaborations with others.

This also leads to another important factor: people's creativity. In the future, we might have more competition when generating innovations. Hence, we need more creative lateral thinkers – and not just for economic reasons. We need these people because the complexity of solving (global) problems is growing. That's also a reason why tomorrow's educational systems might need a stronger integration of creative disciplines. According to creativity expert Sir Ken Robinson, creativity needs to be an integral part of how people are educated: "Creativity now is as important in education as literacy, and we should treat it with the same status" (Robinson 2006). However, Robinson (2006) states, creativity is hard to teach in schools because of the learning environment. There are too many hierarchical structures and the different subjects are taught in isolation from each other and without many interconnections. "Every education system on Earth has the same hierarchy of subjects: at the top are mathematics and languages, then the humanities, and the bottom are the arts" (Robinson 2006). FabLabs encompass basic features of how to practice education in a much more constructive way. People do not only learn how to use high tech fabrication tools, they learn how to pick up knowledge from many different disciplines and share it with others. They learn to bring about their own projects and to help others to accomplish the same. I guess, in 20 years, FabLabs will have provided lots of ideas as how to establish new ways of learning and they will still be an important interface between technology, arts and society.

Up to now I did quite a lot of advertising, including conversations with many institutions and networking with the international FabLab community to support the building of a FabLab in Bremen: Furthermore, I have supported several events, such as FabLab BarCamp, DIY-Lounge, Digital Skylines and fab*education.

FabLab BarCamp

In February 2010, a friend of mine had the idea to organize a BarCamp in Bremen for people who would be interested in the topic of digital manufacturing. We expected some friends and people from the local creative industry to attend the event. After posting the event online and inviting some Dutch FabLab managers to talk about their experience, the event was sold out two weeks later. People from all over Germany came to join us. Most of the guests met each other for the first time. We realized that this was the first meeting of a German FabLab community, which had obviously already existed.

DIY-Lounge

This project was organized together with the 'Sportgarten Bremen e.V.' (an initiative for the youth with an own skate park as well as other leisure activities) and the mobile FabLab truck from Jaap Vermaas. It was designed as an experiment, to figure out how collaboration between a 'trend sport culture' and a 'fabbing culture' would work. The beautiful slogan "dream – make – share" from FabLab Luzern inspired me to do this, because I feel that skaters already practice this idea. They dream about what they can do with their board and immediately try things out and they also share experiences peer-to-peer. In March 2011, we built our DIY-Lounge during the Passions Sports Convention at 'Messe Bremen' (a trade fair). We spent two extremely intense and noisy days together. We found many similarities in both cultures and lots of interfaces to continue with the collaboration.

Digital Skylines

The goal of this format (the project is still active) is to develop continuous collaborations between partners within the metropolitan area of Bremen-Oldenburg-Groningen. First of all, it is designed to exchange ideas and expertise by doing hands-on workshops. Part of this project is also to mix partners from cultural, scientific and educational fields to support networking within the region. At the end of 2011, a group from University of Bremen called dimeb visited the FabLab Groningen to build a 3D printer from Makerbot. In March 2012, staff members and partners from FabLab Groningen came to visit Bremen, to learn about EduWear kits (kits for making interactive and 'intelligent' clothes) and to use Arduino micro controllers in workshops with school classes. Partners on this project were science-linX from Rijksuniversiteit Groningen (a science center of the faculty of Mathematics and Natural Sciences at the University of Groningen), Kunsthalle Bremen (Art Museum Bremen), Focke Museum (Local History Museum in Bremen) and Wirtschaftsförderung Bremen GmbH (an initiative for the promotion of local business development).

Fab*Education

This event was organized to display and analyze the potential of FabLabs in the field of education and to find young talents within 'STEM disciplines'. The challenge of organizing this event was to find a good balance between arranging time for symposiums, exhibitions and workshops. A special focus for me was to discuss the role of the creative industry within this context, especially reflecting on hands-on workshops that we conducted with young people. We received lots of support from the international FabLab community and from our partners in Bremen.

During fab*education, a primary schoolgirl visited the event with her father. Both spent nine hours looking around the exhibition and engaging with the temporary FabLab that we had built. For her, not only the art pieces and machines were fascinating but also the atmosphere, finding lots of open-minded people who could provide answers to all her questions (which were quite a lot). Proudly, she printed her own individual toothbrush tumbler with the Ultimaker 3D printer. In the end, the positive feedback of that girl encouraged the whole team to continue the journey of establishing a permanent FabLab in Bremen.

Reference

Robinson, K 2006, *Schools Kill Creativity*, TEDTalk, viewed 2 January 2013, <http://www.ted.com/speakers/sir_ken_robinson.html>.

FABLABS – A GLOBAL SOCIAL MOVEMENT?

JULIA WALTER-HERRMANN

Never doubt that a small group of thoughtful,
committed citizens can change the world.
Indeed, it is the only thing that ever has[1].
Margaret Mead

1. Introduction

Hardly any debate understands FabLabs only as a "collection of commercially available machines and parts lined by software and processes [...] developed for making things" (Gershenfeld 2005, p. 12) that is assembled in small-scale workshops; most of the currently growing number of texts, presentations and websites about FabLabs also describe it as a 'movement' (e.g. Turner 2010, p. 28; Chugh 2011).

This subsumption is first based on the mere fact that at this stage, a great quantity of approximately 120 FabLabs exists around the globe (MIT FabCentral 2012). Second, from the beginning, Neil Gershenfeld, the initiator and intellectual father of FabLabs, intended not only to "create 'almost anything'" (Nunez 2010, p. 24; his emphasis), but to make almost anything accessible for anybody, too. The notion "that Fab will inspire more people to start their own technological futures" (Gershenfeld 2005, p. 17) does not only focus on the supposedly 'neutral' idea of producing things on one's own, but it is tightly interlinked with other social factors, e.g. the idea of a redistribution of goods and opportunities, a free flow of ideas, documentations and 3D models, and Open Source hard- and software. To sum it up, 'sharing' is an essential aspect of FabLabs. In these sharing processes, the communities of fabbers and FabLabs around the world are the addressees, using the World Wide Web as their communication platform.

This text raises the question whether a network of fabbers and FabLabs with collective ideas, activities and communication channels can be called a global social

1 | This quote is widely attributed to Margaret Mead, a cultural anthropologist and follower of cultural relativism, but at the same time no reliable source is known for reference.

movement[2] or whether FabLabs are better described as a network or simply an accumulation of high-tech workshops. Based on relevant literature, this text asks on a theoretic level what is special about the FabLab movement and how it can be understood nowadays. Hence, based on a social sciences perspective, the text briefly introduces the characteristics of a social movement as an entity with core values, aims for social change and a certain repertoire for realizing this change (Tilly 2004). Therefore, this chapter differs from many other texts about the FabLab movement that rather focus on business models (Gjengedahl 2006) or innovation (Troxler 2010). The contribution then follows up with an empirical description of the worldwide network of FabLabs. Categories that are especially relevant for the empirical analysis are the character and form of the FabLab network and the relation between the several FabLabs, where particularly the role of FabLabs in the "global south" (Harvey 2006) is taken into account. Further categories focus on the individual FabLabs in more detail, i.e. on their equipment, economical infrastructure, organization, values, access and the things that are produced there. These categories are carved out to enlighten the characteristic(s) of the FabLab movement. In the end, the theoretical and the empirical analysis are brought together.

2. People, Social Movements and FabLabs

The Wikipedia entry List of Social Movements mentions Guerilla Gardening, Occupy Movement, Health at Every Size, Arab Spring, Slow Food, Arts and Crafts Movement, Free Data or Nudism as social movements[3] There seems to be an inflationary usage of the word (Tilly 2004, p. 7), yet there is hardly any sharp definition of the term. Christiansen (2009) states that it is exactly the blurriness of the term that is congruent with the blurriness of the discussed topic. Social movements are neither institutionalized parties nor a "mass trend [...] without goals" (Christiansen 2009, p. 2), but social movements can be placed somewhere in the middle of these antipoles (Christiansen 2009, p. 2).

Nevertheless, besides such a vague definition, there still is a more accurate, general definition of what a social movement is, which I depict in this chapter. Before theoretically discussing whether FabLabs can be understood as social movements, I am first going to introduce the initial occurrence of a social movement in Europe and I am also going to briefly explain the history of social movement research. I will then present relevant aspects for the development of a social movement, such as the individual and society on a global scale. I will conclude with yet another theoretical discussion that tries to understand FabLabs as a social movement in the current globalized world.

2 | In this text, all movements referred to are social movements because they are all based on human (social) action, independent from their focus, such as politics, human rights etc.. The usage of the term only aims at distinguishing movements from other forms of action, "such as electoral campaigns, patriotic celebrations, displays of military force, investitures of public officials, and collective mourning" (Tilly 2004, p. 3).

3 | For more information about these movements see the Wikipedia entry List of Social Movements, last viewed 10 January 2013 <http://en.wikipedia.org/wiki/List_of_social_movements>.

2.1 What Is a Social Movement?

Although social movements have a 'long-standing tradition' in Western history, at least back until the appearance of the movements linked to the French Revolution, the academic research on social movements is a relatively new area of research to the social sciences. Despite the fact that the German sociologist Lorenz von Stein first introduced the word 'social movement' in his book History of the French Social Movement from 1789 to the Present in as early as 1850 (Tilly 2004, pp. 4-5), the empirical analysis and understanding of movements, its forces, and counterforces seemed to be negligible for a long time. Movements were either understood as advice-givers for political developments (Tilly 2004, p. 6) or were referred to be chaotic and a danger to social systems. An ambition to scientifically understand the motors, phases, and organizational structures of social movements did not arise until the 1960s and 1970s (Christiansen 2009, p. 2), when social activism influenced whole generations and not only particular social groups.

Social movement theorist Tilly (2004) argues that there are three historically continuing characteristics shared by all movements, which are independent of a movement's goals: These supertemporal factors are "1) campaigns of collective claims on target authorities; 2) an array of claim-making performances including special-purpose associations, public meetings, media statements, and demonstrations; 3) public representations of the cause's *w*orthiness, *u*nity, *n*umbers, and *c*ommitment" (Tilly 2004, p. 7; my emphasis), so called WUNC displays. Tilly furthermore argues that movements "depend heavily on political entrepreneurs for their scale, durability, and effectiveness" (Tilly 2004, p. 13). In this context, the existence of a so-called 'charismatic leader' is often described as essential for the success of a movement. The question arising from such a seemingly political definition is whether FabLabs comply with it.

2.2 Are FabLabs a Social Movement?

For FabLabs, the first characteristic of a movement, the "organized public effort making collective claims" (Tilly 2004, p. 3), can be seen in the 'FabLab idea' written down in the Fab Charter (Fab Charter 2012) that is not only on public display on the Internet, but also in many FabLabs. The second point, the "social movement repertoire" (Tilly 2004, p. 3) of FabLab activities include conferences, fairs, demonstrations and workshops, but also the Center for Bits and Atoms' course on How To Make (almost) Anything or the FabFi project (FabFi 2012). The continuous inclusion of such activities under a collective claim, in this case the FabLab idea, is also typical for a movement (Tilly 2004, p. 3). The third distinctive feature of movements, the WUNC display, can be understood as the corporate identity or the "collective self-representation[]" (Tilly 2004, p. 4) of FabLabs demonstrating values, messages, uniqueness of the cause and their members etc.. Exemplary features of the FabLab's WUNC display can be the representation of 'unique', 'hip' and 'life-enriching' projects at the Fab Academy's website; quotes associated with FabLabs, such as 'The Future is Fab' (laser-cut lettering displayed in the FabLab St. Pauli) or 'Make Things not War' (portrayed on a poster depicting the Afghan FabLab); and the FabLab logo that represents the FabLab idea.

Figure 1: Laser-cut FabLab logos and designs (Source: Design by watsdesign, Thingiverse; this file is licensed under the public domain).

The logo's three-dimensionality refers to the making of real, material, 3D objects, the symmetrical and precise cut outs refer to the tools used in FabLabs (e.g., a laser cutter) and the direction of the cutouts as well as the surrounding circle refer to the circularity of product cycles and a general sustainability of production in the labs. Most FabLabs do not only refer to the Fab Charter, but also to Neil Gershenfeld as their founding father (on the labs' websites or in texts they produce). Apart from Neil Gershenfeld, there are numerous other 'political entrepreneurs' dedicated to the FabLab idea around the globe, the International Fab Lab Association can give an overview of them and their activities (International Fab Lab Association 2012). Based on the analysis of distinct features of social movements in general, it can be said that FabLabs are in fact a social movement. But although there are features shared by each movement, social movements are no "phenomenon sui generis" (Tilly 2004, p. 9), they have to be understood in their historically and socially specific context, which is why I am now going to introduce an approach of social movements trying to understand a social movement based on the character of the society and the individuals it belongs to (Tilly 2004; Kern 2008, pp. 189 et sqq.).

2.3 Movements and the People, Now and Then

Oftentimes, authors refer to Karl Marx in order to explain and further describe FabLab ideas and practices (e.g., Blanc 2012; Boeing 2008; Fischbach 2008). The combination is obvious: Karl Marx is the "key philosopher [and critic] of industrialism" (Touraine 2004, p. 129), whereas at the same time, FabLabs reference a pre-industrialized past. A time "before art was separated from artisans, when production was done for individuals rather than the masses" (Gershenfeld 2005, p. 8), when producers were not yet alienated from their products and the means of production were not only used by laborers, but also owned by them (Marx 1932). Gershenfeld even states, that "possession of the means for industrial production has long been the dividing line between workers and owners […] That boundary will recede until today's marketplace evolves into a continuum from creators to consumers" (Gershenfeld 2005, p. 15). FabLabs virtually manifest Marx's vision of society as "an association of free people that work with common means of production and that give their individual labor self-confident to a community" (Marx 1932, p. 91; own translation). Even though the actual operational procedures in a FabLab might be comparable to what Marx had in mind for an 'ideal society', the movement of FabLabs cannot be explained by Marx's ideas, because the individuals and the society that bring up a movement are not comparable to those of Marx's times. Furthermore, FabLabs or 3D printers currently cannot induce Marx's vision of an 'ideal society', like some authors propose, because the scale of production in FabLabs as well as the available technologies are not (yet) adequate for redistributing the means of production.

In central Europe of the late 19th century, the significant conflict line was drawn between the lower and the upper socio-economical stratum, between the workers who produced most of the goods but owned the fewest, and the people who owned and consumed most of the goods. The bisection of society was rather simple. Being part of the same system – "a civilization of labour" (Touraine 2004, p. 718) – the lower class easily knew whom to blame for their situation and whom to fight and what to fight for. For the proletarians a conflict arose, as they could not participate in the (capitalist) system, leading them to form a collective identity, the labor movement becoming their mouthpiece (Kern 2008, p. 193).

In order to further understand FabLabs – which are scattered around the globe – as a global social movement, one has to put the frame of analysis from 19th century central Europe to the present globalized world. It can be argued that the distribution of goods around the globe has some similarities to those in the 19th century, i. e. comparatively few people own almost everything today (Harvey 2006, pp. 95-96). But at the same time, 'the world' can hardly be understood as one system[4], and its inhabitants do not only belong to two groups within this system, like proletarians and capitalists. An increasing organization and functional differentiation of function systems into sub-systems can currently be observed worldwide (Kern 2008, p. 191). "Organization and functional differentiation are considered key principles of modern societies [...] Autonomous function systems [can be, for example,] politics, the economy, science, art, religion, etc." (Roth 2013). An organization and functional differentiation of the science system could be the differentiation into various scientific subjects, disciplines, forms of education, theoretical approaches and so forth continuously dividing the science system into smaller sections. In addition to the further differentiation of the systems, an increasing individualization of people can also be observed (Kern 2008, p. 194; Beck 1986). Individualization means that attributions of people to a certain class become obsolete and an increasing pressure for a reflexive and self-determined lifestyle exists, which leads to a pluralization of lifestyles (Beck 1986). These key principles can be witnessed for all modern societies, irrespective of their status as high- or low-income countries. The notion of an uneven global development rather supports the idea of a differentiation of systems (Harvey 2006). Functional differentiation and individualization reside in a relation of reciprocal cumulation (Kern 2008, p. 192); the more the sub-systems are differentiated the more the individuals can participate in their very specific sub-systems, which then again

4 | Still, for analyzing the world it can be theoretically thought as one system. This is what lots of social scientists actually do, e.g., Luhmann, Castells, Meckel and many others.

leads to further individualization; and the more individualized the people are, the more specific the functional systems have to be to fulfill the individuals' very specific needs. A conflict always arises when an individual cannot participate in a (sub-)system (Kern 2008, p. 192) – like the conflict that arose for the proletarians who could not participate in the capitalist system of the 19th century. Due to the systems' virtual infinite expansion, such a conflict only arises if a) a subsystem's capacities are pushed beyond its boundaries and b) if the ecological resources draw to a close. A movement calls attention to such conflicts and claims or tries to establish the participation of 'affected' individuals in the system (Kern 2008, p. 192).

Several questions occur from the description of the current social system and its individuals concerning the characteristics of current movements, such as FabLabs. 1) How do movements appear in a world that is highly differentiated into various sub-systems, where traditional business models and capitalist economies are shaken up, where people are highly individualized, belonging to particular groups, having very special needs? 2) What are their conflicts? 3) How are these conflicts expressed? And 4) how and why are people even mobilized if there is practically always a possibility of avoiding a conflict by extending a sub-system or 'escaping' into another?

2.4 FabLabs – A Social Movement of the Present Age

In this passage, I theoretically discuss the status and the characteristics of the FabLab movement, based on the description of the current global world and the questions that were derived from that description. 1) Social movements that occurred in modern societies since the 1960s, as described before, are often described as 'new social movements' (Kern 2008, p. 189). These movements are detached from state affairs and mainly reside in the cultural or social sphere. They reject materialistic goals, in contrast to early labor movements, and replace these goals with rather subjective, immaterial and cultural values that have a positive connotation, like creativity, free sexual orientation or a moral attitude towards nature and its preservation (Castells 2004; Touraine 2004a). Therefore, the current movements are tightly linked to individual conflicts, thus being highly individualized movements – which can be the same for people all over the world – like Guerilla Gardening, Occupy Movement or FabLabs. In any case, a movement revolving around high-tech production machines in small workshops was probably rather unthinkable some years ago. 2) The conflicts negotiated through movements are as different as the individuals having them, ranging from not being able to take part in the design of public spaces, not being able to take part in a socially fair distribution of goods, or not being able to produce (almost) anything. Since all these claims are actually rooted within the individual sphere and are based on an individually important conflict about not being able to take part in a system, a judgment about the importance or worthiness of a claim expressed through a movement is nearly unfeasible (Rucht & Neidhardt 2007, p. 3). What is furthermore typical of current movements is that they do not need to address a certain 'oppressor', who is not easily identifiable in a world consisting of extremely differentiated functioning systems, anyway. Current movements either address their aim to a general public or try to achieve the desired change for themselves. FabLabs even incorporate the latter aspect and make it part of their claim: 'making', which could also be called 'self-making', is an essential aspect of every FabLab. Furthermore, today's movements express their conflicts in various ways, not only through walkouts and demonstrations like in the past, but also through media campaigns, establishing Twitter Hashtags, organizing workshops, producing videos ... the list of (creative) actions could be continued endlessly (Rucht et al. 2004). By making stuff, people in FabLabs do not only solve the conflict they claim, but also express the FabLab idea through their making and exhibiting machines and objects on the Internet, at fairs or in show cases. And lastly 4), literature says, because people experience fewer conflicts, they are in fact less motivated to engage in movements, but do more often engage in secluded self-expression and personal fulfillment (Touraine 2002, pp. 390 et sqq.; Kern 2008, pp. 189 et sqq.). At the same time, in an individualized and functional differentiated world, a WUNC display regains importance. Since people are seeking individuality and very specific sub-systems, a movement can work as their relevant system, whereas a movement's WUNC display offers an adoptable identity, with very specific, extraordinary attributes, such as special clothes, shared symbols and languages. In a workshop conducted by Corinne Büching and me in the FabLab St. Pauli (see Büching in this book), we handed out questionnaires to the participants. Though there was no question about the image the participants have of FabLabs (only a blank box for comments), eight out of eleven participants used the blank to write about the impression they have of FabLabs: 'innovative', 'hip', 'fun', 'highly interesting' and 'super cool' were buzzwords the participants used to describe FabLabs (Walter-Herrmann & Büching 2012; Walter-Herrmann 2012). Based on these findings, it can be said that the very specificity of a FabLab as a functioning system as well as a FabLab's image as an innovative, open space, where everybody can be an artist, and the adoption of these characteristics into one's own identity, can be factors that motivate people to participate in the movement. In addition to the 'hip image' of FabLabs, they are relatively open in the image they represent and individual FabLabs and people can attribute their own claims to it. Consequently, people can identify with the cultural, artistic or even materialistic approaches of FabLabs.

In this chapter, I first introduced the characteristics of social movements and its connection to society. I theoretically analyzed FabLabs as movements of our current, individualized, and functionally differentiated world. I further carved out that FabLabs are an individualized, unique movement that takes 'making' as an essential claim, as well as a part of its "social movement repertoire" (Tilly 2004, p. 3) and whose image is important for its members' motivation.

3. Empirical Methods and Analytical Categories

A theoretical analysis of FabLabs as a global social movement can help understanding FabLabs in the current world and its role for society. Due to such a theoretical examination certain policies and patterns, which previously might have been understood as typical and unique for FabLabs, can thus be carved out not only as typical for FabLabs but for movements in general and therefore better understood.

Gary Marx (1979) claims not only external factors for the success of a movement as relevant, such as conflicts with a current system and the state of society, but also internal factors, that are not only found on an individual level. Internal factors can be the people's collaborations or the organizational models and information policies of movements (for more information about the organizational infrastructure of FabLabs see Troxler in this book). An analysis on a theoretic level can never grasp such internal factors, because these factors are mostly inaccessible for an outside researcher, unless they are made public. The weakness of such analyses is their abidance on the surface of 'logical assumptions' about a certain topic that is visible from outside.

To further the understanding of FabLabs as a global social movement I am now going to present an empirical study about the nature of FabLabs as a global social movement. I will first introduce the methods I used to research the movement of FabLabs and critically reflect my method of inquiry before presenting my findings and relevant categories.

3.1 Empirical Methods

My aim for researching the FabLab movement was to ascertain more knowledge about the way individual FabLabs are linked to each other, work together, and communicate, especially concerning the relationship between the FabLabs of the Western world and those of the global south. In addition I wanted to know what kind of people come to the individual labs, what a lab's specific FabLab idea is, what its claim is and how the labs are financed. My questions allow the deduction, that I was not interested in the internal factors of a 'FabLab's grassroots', such as visitors, students, voluntary helpers or members, their motivation for taking part in the movement or making things, but in the internal factor that each FabLab contributes to the global movement. Therefore – after making a pre test with colleagues from my working group – I e-mailed standardized quantitative questionnaires with 25 multiple choice questions, each with three answering options[5] and a blank box (survey methodology) to all FabLabs that were listed at the Center for Bits and Atoms' website (Center for Bits and Atoms 2012), which was around 110 at that time. My hope was – without knowing who will open the mail – that a person in charge would be responsible for the mails sent to a lab's general address; a person that would be capable of telling me about the insides in the FabLab's structure and claims. The questionnaire can be understood as a so-called expert survey. Not all FabLabs' mail addresses were correct, so in the end I send little less than 100 inquiries to the FabLabs. After four weeks only 16 labs had answered the questionnaire, so that I asked a FabLab-activist from Germany, who is deeply involved in the FabLab community, to announce my request at a general mailing list – without personally asking friends to answer it for not biasing the answers to a German or Europe perspective. After two additional weeks 21 people had answered the questionnaire. Though it can be said the questionnaire is not at all

5 | The response format is based on the so-called 'nominal-polytomous', where a question can be answered with more than two unordered options. In my case three unordered options were possible, but voluntary. Based on this response format a total response rate of 300% for all answers and 100% per a single answer is possible.

representative, a response rate of about 20% is, statistically speaking, neither unusual nor an indication of low accuracy of a survey (Holbrock, Allyson & Pfent 2007, pp. 499 et sqq.). Usually a low response rate is balanced out by increasing the population, which is not possible for FabLabs, because its population is only 120. Anyway I do not like to pretend the categories and findings I could establish are significant for all FabLabs, but maybe they can give interesting hints towards certain developments and are worth discussing.

3.2 Empirical Findings about the FabLab Movement

As this chapter is entitled FabLabs – A Global Social Movement? the theoretical analysis could already reveal that FabLabs are a global social movement, now I am going to describe how this global movement is organized, structured and what its claim is. Based on the quantitative answers to the questionnaire the following relevant categories could be grouped: 'The FabLab Network', including communication channels and some aspects of finance, 'The FabLab Idea', including objects produced in the respective FabLab and its users and 'The Individual Labs', including forms of organization and finances.

3.2.1 The FabLab Network

My definition of network entails the sum of FabLabs and the relationship(s) they maintain. From the 21 FabLabs, which answered the questionnaire, twelve are based in North America, six in Europe and three in Asia. Regrettably no FabLab from South America or Africa answered the questions, so that I can hardly provide any information about my initial question about the relationship between FabLabs from the Western world and the global south. Nearly all FabLabs stated they had most contact to another FabLab that is located on their continent, which is interesting, because FabLabs are so tightly entangled with digital communication technologies, that one could think 'real distances' should be less important. Two FabLabs from Asia are the exception from this observation: they both named the Center for Bits and Atoms' (CBA) FabLab as the one they had most contact with. One of them also stated, that their relationship is characterized by financial support. So it can be said for these specific FabLabs in Asia the CBA's FabLab plays a more important role than for others though their relationship is not inevitable characterized through a financial dependency. For all other FabLabs the character of the relationship to the FabLab they have the most contact with is in equal parts based on 'personal friendship' and 'project collaboration'. This means, the network between the FabLabs is a) rather informal and b) closely connected to certain individuals. A conclusion that can be backed up by the preferred communication technologies FabLabs use for staying in contact. With nearly 90%[6] each, 'face-to-face communication' and 'e-mail' were the most common answers. For the second question, about naming another lab one FabLab has much contact with, the answers seemed rather random – or at least difficult to interpret. The variety of named FabLabs was much larger and less dependent on real distances than at the first question, though none of the participating FabLabs from North America, Europe and Asia stated they had

6 | All data are rounded.

most contact (in the first or second place) with a FabLab in Africa, Asia or South America. The relationship with another FabLab is best described as an 'exchange of ideas', as about 70% of the participants said. 75% mentioned 'websites and social network sites', but also 'e-mail' (with 48%) as important communication channels for communicating with a FabLab that is also important for them. This means within the FabLab network strong ties (project-collaboration, friendship) as well as lose ties (check another FabLab's website, exchange ideas) are possible and do exist.

3.2.2 The FabLab Idea

By the term FabLab idea I mean the orientation and focus of activities a FabLab has and the claim it is making, which is mirrored in the produced objects and the users who come to a FabLab. As mentioned before the idea itself is already put on record in the Fab Charter (Fab Charter 2012), but I will now describe, whether particular variations from the Fab Charter exist. 20 FabLabs expressed 'to educate' as one of their aims, as it is already written in the Fab Charter. Varying between six and nine nominations 'to empower', 'to make research', 'to offer high-tech technologies', 'to offer leisure times facilities' and 'to work as a design workshop' are comparably relevant aspects for the FabLabs participating in the survey. Besides that, it is noteworthy, though media coverage and various texts often focus on these features, the aspects 'to leverage digital fabrication to sustain a global revolution' and 'to offer the opportunity to satisfy people's needs and shortages' were only named once, as well as the produced objects' characteristic 'to correct a defect or shortage'. Other characteristics of the objects focus around two aspects: 'individually meaningful objects' and 'educationally meaningful objects'. The produced objects' specifications 'singular pieces / unique', 'gadgets / knick-knack' were named as often as 'prototypes', and 'technically complex' objects. Two thirds of the FabLabs specified their users as 'technically interested', just as many said, their users 'have a higher educational attainment' and each with a 50% response rate claimed their users 'work in the creative sector' or 'work in the education sector', they are either 'organized groups' and 'individuals with a concrete idea' or 'students who come to the lab as part of their curricula'. To sum this up, all FabLabs follow an educational approach that mostly attracts well educated, technology interested people, but also a further distinction between FabLabs as educational and design spaces can be conveyed from the descriptions of objects and users. Presumably FabLabs especially fill a demand for accessible spaces for tinkering with technology, like gyms for sport and rehearsal rooms for music (for more information about valued spaces see Blikstein in this book). World-revolutionary approaches play a rather inferior role for FabLabs. Due to aspects of objectivity the survey only asked for actual users who come to a lab and not for intended target audiences, though if they were not congruent FabLabs should overthink their WUNC displays.

3.2.3 The Individual Labs

The description of individual labs is directly connected to the FabLab idea of the labs: a distinction between those labs that 'belong to an educational institution' and those FabLabs that 'evolve around communities and associations' can be made, as both answering options were named with 70% each. Anyway all participating FabLabs seem to be based on the same financial model: 'get subsidies from the state and / or any public-law institution' was the most relevant answer with about 90% of answers, 'impose fees for courses, workshops and /or materials' was the second most relevant response, with 70% of the participants answering in that way. All FabLabs stated they have less than $5000 average disposal per month and all spend it on a monthly basis mostly for 'room rent / electricity / Internet access / other additional expenses etc.', 'machines / computers / books / software etc.' (with a 70% response rate each) and 'regular staff's payment' (covering 50% of answers). Since sustaining business models should not include one-off costs within their regular expenses, the answers are interesting to that effect that the biggest costs are simultaneously one-off and regular costs. It can also be possible that the survey participants either allocated the high acquisition costs of FabLabs to a monthly basis or misunderstood the question. It is also thought-provoking that only few FabLabs seem to be able to employ regular staff to maintain their labs.

4. Conclusion

Based on texts that have called FabLabs a movement, this chapter raised the question whether FabLabs are a global social movement. Therefore it introduced features of global movements, which are either general or historically specific. It was discussed and confirmed on a theoretical basis, that FabLabs fulfill this definition. Furthermore, the FabLab movement was also analyzed empirically, based on the survey methodology, aiming at detecting internal features of the movement. Though the findings are not representative, some surprising aspects could be described. For example, the literature almost always mentions that singular conflict, claim or orientation a movement has to have. Contrasting to this, the empirical description showed that (throughout the FabLabs participating in the movement) two foci can be identified: the integration in educational institutions, with a focus on curricular education, with students being the main audiences, and community driven design spaces, with individuals as main audiences. Furthermore, the social science literature about social movements hardly ever speaks about internal structures of movements, but the FabLab survey showed that weak and strong ties can be included within one movement. Moreover the theoretical as well as the empirical analyzes showed that an understanding of FabLabs as 'revolutionary places for class struggle' is a rather inferior aspect for FabLabs. Eligible follow up questions are now: can these first findings be confirmed on a representative basis, how is the concrete relationship between the FabLabs of Western world and the global south and what is the relationship between a FabLab's individual WUNC display and its actual FabLab idea.

References

Beck, U 1986, *Risikogesellschaft. Auf dem Weg in eine andere Moderne*, Suhrkamp, Frankfurt, Germany.

Blanc, S 2012, '3-D-Drucker mit Schokopatrone: Eine Software, die alles baut, was Sie brauchen' in *Le Monde Diplomatique*, no 9821, viewed 10 January 2013, <http://www.monde-diplomatique.de/pm/2012/06/08.mondeText.artikel,a0062.idx,23>.

Boeing, N 2008, Die Marx-Maschine, media release, 29 February, Der Freitag, viewed 10 January 2013, <http://www.freitag.de/autoren/der-freitag/die-marx-maschine>.

Castells, M 2004, *The Power of Identity*, Blackwell, London, GB.

Christiansen, J 2009, 'Four Stages of Social Movements', in *EBSCO Research Starters*, EBSCO Publishing, viewed 10 January 2013, <http://www.ebscohost.com/uploads/imported/thisTopic-dbTopic-1248.pdf>.

Chug, A 2011, *FabLab Links*, viewed 11 January 2013, <https://sites.google.com/site/fablablinks/the-network>.

Fab Charter 2012, viewed 11 January 2013, <http://fab.cba.mit.edu/about/charter/>.

FabFi 2012, viewed 10 September 2012, <http://fabfi.fabfolk.com/>.

Fischbach, R 2008, Die Marx-Maschine? Die Murksmaschine, media release, 20 March, Der Freitag, viewed 10 January 2013, <http://www.freitag.de/autoren/der-freitag/marx-maschine-murksmaschine>.

Gjengedal, A 2006, 'Industrial clusters and establishment of MIT FabLab at Furuflaten, Norway' Proceedings of *The 9th International Conference on Engineering Education*, San Juan, Puerto Rico, viewed 10 January 2013, <http://www.ineer.org/Events/ICEE2006/papers/3600.pdf>.

Gershenfeld, N 2005, *Fab. The coming revolution on your desktop – from personal computers to personal fabrication*, Basic Books, New York.

International Fab Lab Association, viewed 11 January 2012, <http://fablabinternational.org/>.

Harvey, D 2006, *Spaces of Global Capitalism: A Theory of Uneven Geographical Development*, Verso, London, New York.

Kern, T 2008, *Soziale Bewegungen. Ursachen, Wirkungen, Mechanismen*, Verlag für Sozialwissenschaften, Wiesbaden, Germany.

Marx, G 1979, 'External Efforts to Damage or Facilitate Social Movements: Some Patterns, Explanations, Outcomes and Complications', in Zald, M & McCarthy, J (eds.), *The Dynamics of Social Movements*, Winthrop Publishers, Cambridge, MA.

Marx, K [2009] 1932, *Das Kapital: Kritik der politischen Ökonomie*, [reprint by Anaconda Verlag, Köln, Germany] Kiepenheuer Verlag, Berlin, Germany.

MIT FabCentral 2012, viewed 11 January 2013, <http://fab.cba.mit.edu/about/labs/>.

Nunez, JG 2010, 'Prefab the FabLab: Rethinking the Habitability of a Fabrication Lab by Including Fixture-based Components', Master's thesis, Massachusetts Institute of Technology, Cambridge, MA, USA, viewed 10 January 2012, <http://dspace.mit.edu/handle/1721.1/59201>.

Roth, S 2013, *The Ten Systems: On the Canonization of Function Systems*, Social Science Electronic Publishing, Inc., viewed 10 January 2013, <http://dx.doi.org/10.2139/ssrn.2148935>.

Rucht, D & Neidhardt, F 2007, 'Soziale Bewegungen und soziale Aktionen', in Joas, H (ed.), *Lehrbuch der Soziologie*, Campus, Frankfurt/New York, pp. 533-556.

Rucht, D, van de Donk, W, Loader, B & Nixon, P 2004, *Cyberprotest: New Media, Citizens and Social Movements*, Routledge, London, GB.

Tilly, C 2004, *Social Movements, 1768–2004*, Paradigm Publishers, Boulder, Colorado, USA.

Touraine, A 2002, 'From understanding society to discovering the subject', *Anthropological Theory*, vol. 2, pp. 387-398.

Touraine, A 2004, 'On the Frontier of Social Movements', *Current Sociology*, vol. 52, pp. 717-725.

Touraine, A 2004a, 'Neo-Modern Ecology', *New Perspectives Quarterly*, vol. 21, no. 4, pp. 129-133.

Turner, R 2010, 'Open Source as a Tool for Communal Technology Development: Using Appropriate Technology Criteria to Determine the Impact of Open Source Technologies on Communities as Delivered Through the Massachusetts Institute of Technology Fab Lab Projects', Master's thesis, Wits School of Arts, University of the Witwatersrand, Johannesburg, South Africa.

Troxler, P 2010, 'Commons-based Peer-Production of Physical Goods. Is there Room for a Hybrid Innovation Ecology?', Proceedings of *The 3rd Free Culture Research Conference*, Berlin, Germany, viewed 10 January 2013 <http://wikis.fu-berlin.de/download/attachments/59080767/Troxler-Paper.pdf>.

Walter-Herrmann, J & Büching, C 2012, 'A FabLab Learning Scenario for Young Adults', paper presented at the 11th International Conference on Interaction Design and Children, Bremen, Germany, 12-15 June 2012.

Walter-Herrmann, J 2012, 'What Can We Learn in FabLabs – and how can FabLabs be learned?', paper presented at the FabLearn 2012: Digital Fabrication in Education Conference, Stanford University, Palo Alto, USA, 17-18 October 2012.

HOMO FABBER AND THE LAW[1]

LAMBERT GROSSKOPF

1. Technology and its Applications

3D printers replicate not just themselves but everyday objects as well. Professional 3D printers can be used in 3D print shops, the copy shops of the future, and in social manufacturing services, anyone can offer the capacity of his or her 3D printer, or distribute templates to print utility or design objects. Even The Pirate Bay now includes a Physibles section for 3D models. Online platforms such as Shapeways, Ponoko and i.materialise offer a professional 3D printing service and also provide their users with easily operated online tools for producing and editing design files. The future of 3D printing is already being demonstrated by Teenage Engineering, a Stockholm-based company in Sweden. Instead of sending spare parts by post, customers can download a 3D model and produce the spare part themselves, using their 3D printer.

On their own 3D printers, most prosumers can currently only manufacture somewhat simple utility and design objects made of a single material (plastic, resin, plaster, metal powder, sand or brothy foods). However, experiments are already being conducted with printing techniques involving several materials, or using materials in different colours.

The digital templates are created using CAD (Computer Aided Design) software that can now be operated by non-experts. Either the 3D objects are modelled with the aid of software, or an existing object is photographed from several sides, with a 3D model then being computed using the images obtained. Templates can also be created using 3D scanners, for example with the Kinect motion controller.

1 | This paper describes the legal situation in Germany only.

Professional 3D printers with which large objects can be made cost much more than 100,000 Euros in many cases, although there are also smaller models for home usage, costing less than a thousand Euros on the market. With the latter, however, it is already possible to produce simple household goods, to make parts for hobby crafts, to print toys, or create jewellery. Few constraints are placed on one's own creativity, but not everyone will buy his or her own 3D printer. FabLabs or 3D print shops will come into being instead, similar to conventional copy shops, but allowing various 3D printers to be shared.

2. Laws that the Homo Fabber Should Be Aware of

3D printing raises a plethora of new legal issues, however. If the printed object is identical or similar to a protected product, this may constitute an infringement of existing copyrights, trademark rights, patents, utility models, or registered designs. For example, the Games Workshop Group, a game production company and maker of miniature figures for board games, has already taken action to remove 3D templates of the popular miniature figures from Lord of the Rings and Warhammer from Thingiverse, a website for sharing 3D design files.

2.1 Industrial Property Rights

A patent is granted for an invention that is novel and industrially applicable. It does not provide protection for eternity, but only for a particular duration. When the patent expires, the invention is free to be used. The pealess whistle, for example, was protected by a patent until 1966. Since then, whistles with two chambers tuned to easily distinguished pitches may be manufactured by any company. A utility model is the 'little brother' of a full-blown patent. In contrast to patents, a utility model is not intellectual property that has been subjected to examination but a right that is derived purely from registration. Whether a utility model actually confers protection or not is something that is not generally established until infringement proceedings are conducted. One example of a utility model was a goalkeeper's pullover protected on the outside with padding that was aimed at preventing the goalkeeper from being injured in goalmouth scrambles. A registered design is an industrial property right that grants the holder the exclusive power to exploit an aesthetic design (design, color, shape), for example for a key ring in the shape of stylized football players. Design protection ensues when the design is registered in the Design Register.

If the extent of protection conferred by a patent is violated, the patent proprietor has rights to injunctive relief, damages, information, submission of accounts, and the elimination of disturbance and to claims based on unjust enrichment. The same applies to utility models. A utility model may likewise be used by its proprietor only, and for that reason may not be used by others without the proprietor's consent. The same is true of registered designs. However, the legal effect of a patent, utility model or registered design does not extend to actions performed in the private sphere for non-commercial purposes. Patents, utility models and registered designs may be used to satisfy directly one's own needs, in particular one's domestic needs.

In the private sphere, not only is it permissible to use a protected product or to apply a protected method or process, one is also allowed to produce any number of a protected product for private use, which therefore includes production using 3D printers. Even offering such a product to others or bringing it into circulation is permitted, but only as unpaid neighbourly assistance. However, used products may subsequently be sold if they were initially produced within the private sphere. In addition to 3D products, it is also permitted to pass on 3D templates free within the private sphere, free of patents, utility models and registered designs. However, if a product or template is to be passed on to somebody in a permissible manner, the objective risk of intellectual property being infringed means that the person offering the product or design file must draw attention to the fact that use is only allowed in the private sphere.

Online marketplaces disseminating 3D templates that infringe patents, utility models or registered designs are committing acts of contributory infringement, due to breach of their 'Verkehrspflicht' (i.e., their 'duty to safeguard traffic', or duty of care) by not preventing or averting the infringement of intellectual property – in that such infringement is facilitated by the dissemination of 3D templates with the aid of the infrastructure provided (liability as accomplice). However, online marketplaces are only liable as accomplice, due to breaches of their duty of care, when there are specific indications of infringement or when they have been informed accordingly about breaches of law (with a cease and desist order). There is no general obligation to check every 3D design file on the portal for potential infringement of intellectual property. However, if there are reasons to suspect such infringement, the portal operators must contact the poster for information about the 3D file. If the poster remains silent, the Internet portal operators must conduct their own review, if necessary by consulting the relevant experts.

On-demand manufacturing of 3D products in order to satisfy private needs directly is permissible, since otherwise only those who can afford their own 3D printer would be able to enjoy the privilege. Acts commissioned from on-demand manufacturers without the consent of those holding the rights to patents, utility models and registered designs are not unlawful, therefore. This is conditional, of course, on the on-demand manufacturer not transgressing the limits to that privilege by manufacturing stocks of products for which there is a strong demand, for example, or by storing 3D design files in order to make them available to third parties. Commissioning an on-demand manufacturer with a 3D printer to make products that make use of patents, utility models or registered designs is permissible even when such production is not free and goes beyond reimbursement of pure expenses or the wages paid to the employee entrusted with the task of making the 3D product.

2.2 Trademark Law

According to the German Trademark Act ('Markengesetz'), third parties not having the consent of the trademark proprietor are prohibited from using signs for products or services that are identical or similar to those protected by the trademark. The same applies to business names and statements of geographical origin that are used in the course of trade to designate goods or services. Trademark law does not apply to private activities, however, because the prerequisite for an infringement of trademark rights is an act committed in the course of trade. A mark is used in the course of trade when use occurs in the context of a commercial activity pursued for financial gain, and not in the private sphere. This means that goods bearing a trademark may be used as a specimen for producing one's own 3D products, or 3D design files, for directly satisfying one's own needs. 2D or 3D trademarks may also be replicated in order to serve one's own needs. A private individual who dresses himself or herself in clothes bearing the trademark is not infringing the trademark in any legally relevant manner.

Online marketplaces that disseminate 3D products or templates on which trademarks are depicted, or which represent 2D or 3D trademarks, are not liable as tortfeasors or as accomplices to the infringement of rights if they have no knowledge of the specific infringement of rights that threatens to occur on their online marketplace. Nor are they liable as accomplices due to a failure to do something, in breach of their obligations. However, the operators of the online marketplace do bear liability if the platform user is a person engaged in a trade or business, but only when it is clearly evident to the operator of the online marketplace that the person is in fact conducting a trade. Whether this is evident or not may be based on circumstances other than the actual offer of products, such as recurrent presence of the seller, or recurrent offering of the same kind of 3D products or 3D design files.

There is no general obligation to perform prior inspections, however. Nor is the online marketplace obligated to check content at a later point, unless there is a specific reason to do so. However, the operator has a special duty to check content if he has already been informed of at least one infringement of rights of some significance and there is a manifest risk of further infringements of rights by individual users.

On-demand manufacturing of 3D products, on which trademarks are shown or which represent 2D or 3D trademarks, is permissible if application of the trademark serves to directly satisfy private needs and as long as there is no perceptible evidence for the on-demand manufacturer that the 3D products made in on-demand manufacturing are being placed on the market.

2.3 Copyright Law

Copyright law safeguards exclusive rights to exploit and to prohibit the use of created works. The extensive rights that are granted by copyright are not unlimited, however, but are confined to certain kinds of use in order to protect freedom of opinion, freedom of the press, broadcasting freedom, freedom of the arts, freedom of science and freedom of information, and to protect the private sphere. In addition to 3D products, 3D templates may also be produced and passed on to others within the private sphere, provided that they are copied and passed on within the private sphere for non-commercial purposes. A design does not enjoy protection as intellectual property if its features are based purely on technical factors, even if they are freely selectable or replaceable, and there is no evidence of artistic performance. A work does not get copyrighted simply because an object has been designed or crafted or one technical feature has been exchanged for another.

A work is deemed to be copied or reproduced not only by one-to-one copies, but also when the work is transformed into a different material, a different dimension, or a different size. Not only is the reproduction of 2D or 3D works in the form of photographs or films deemed to be copying, but also the execution of plans and drafts in respect of such works. The manufacturing of 3D products from creative works therefore constitutes copying as well, because the only relevant criterion is whether or not the embodiment reproduces the work as such. The material used – be it plastic, resin, plaster or metal powder, sand or potato starch – is of no legal significance. The same applies to the production of 3D design files from creative works, which has already led to the first legal dispute over copyright. The case in question concerned a '3D Penrose triangle', an impossible figure. Impossible figures are optical illusions in the form of 2D figures representing 3D constructs that cannot physically exist. A designer published a photo of such an impossible figure. Another designer developed a 3D design file for printing the 3D Penrose Triangle and published the design file on an online platform for digital designs, which promptly received the first ever takedown notice for a 3D design. Since a work is protected not only as a whole, but also in its separate parts, using only parts or individual elements of a work copyrighted by someone else is also deemed to be copying or reproducing in violation of copyright. However, it is then necessary to examine whether the parts being used can be protected singularly.

The act of copying or reproducing a work does not necessarily need to be done by a private individual, but can also be performed by a third party (manufacturing on demand); however, the actions of the third party must be limited to the technical process of reproduction, and the third party must adhere to specific instructions for making the copies. It is not generally permissible to produce an unlimited number of copies, however. The relevant criterion is the respective purpose being pursued in making the copies. In one particular case it may only be permissible to make one or two copies, whereas several copies may be made in a different case. The decisive aspect in the last analysis is how many copies are needed to cover one's individual requirement.

The creators or authors of works have the exclusive right to make their works available to members of the public from places and at times of their choosing (Internet law). 3D design files of copyrighted works that are uploaded to online marketplaces and discussion forums in the public domain, or disseminated via file sharing systems are included in that law. Providing links to content is not deemed to be making something publicly available, in contrast, because it is not the link that makes the work publicly available, but the person who puts the work on the Internet. RSS feeds are not affected, either, because these are push rather than download services.

3. Outlook

When engaging in 3D printing, a consumer departs from his or her traditional role to become a Homo Fabber. Consumers become producers of goods and customise goods to satisfy their own wishes. They do not come into conflict with statutory IPRs and copyright as long as they make 3D products only for themselves or friends and relatives. However, if one-person factories start to multiply, this may well have impacts on the sales revenues of the conventional consumer goods industry. There would then be a risk that the 'old industries', anxious to preserve their markets, cry out to lawmakers to help protect their traditional business models, just like publishing houses or the music industry in recent years. There is therefore cause for concern that either industrial property rights and copyrights may no longer be freely used to satisfy individual needs directly, specifically domestic needs, or that the use of 3D printers in FabLabs or 3D print shops will be prohibited, in other words that shared use of 3D printers will be rendered impossible. In order to produce increasingly sophisticated 3D models himself or herself, Homo Fabber must therefore keep in mind not only technical trends, but also the applicable legislation, if a court judge shall not become the production manager of his or her basement hobby room.

GENDERED FABLABS?

TANJA CARSTENSEN

1. Introduction

Traditionally, technology has been linked with power and masculinity (Cockburn 1986; Wajcman 1991). Even now, women continue to be underrepresented in technical jobs and education. Prejudices about male technological skills and female technological ignorance are widespread. The reasons for this are complex. On the one hand, the emergence of the job profile of the engineer in the 19th century was tightly connected to discourses about hegemonic masculinity; these discourses persist in the 21st century (Zachmann 2004; Paulitz 2012). On the other hand, gender stereotypes are still reproduced by socialization: stereotypical toys, a lack of role models in kindergarten and school and examples in schoolbooks reinforce the idea of technology as a field for boys and men (Knoll & Ratzer 2010). In the media – for example, soap operas – technologically competent female role models are rare, too. And last but not least, technological spaces such as technical universities or companies are still male-dominated and often have developed a specific male culture (Wajcman 1991).

Figure 1: Vinyl cut ornament (Source: Photography by Axel Sylvester).

At the same time, the narrow relation between technology and masculinity does not only refer to questions of access and the exclusion of women from technology, female underrepresentation in technology and technology as a male culture. Gender and Technology Studies have also emphasized that gender relations have an impact on construction processes and the way technological artifacts are designed. Designers (unconsciously) inscribe different views of female and male users and uses into technology, and by this they reproduce their ideas of gender relations. Gender is imprinted onto objects through instructions, advertisements, associations with gendered divisions of labor and associations with gender symbols and myths. Artifacts that incorporate a so-called 'gender script' then contain materialized gender stereotypes and contribute to the construction of users' gender identities (Berg & Lie 1993; van Oost 1995; Cockburn & Ormrod 1993; Oudshoorn, Saetnan & Lie 2002; Zorn et al. 2007). They are therefore powerful, materialized co-players in gender relations (Haraway 1991, p. 153).

Cockburn & Ormrod (1993) show that the question of whether or not an artifact is considered a technology can also change. Using the example of the microwave, they carve out how this initially male-connoted artifact that targeted bachelors evolved into a female-connoted kitchen appliance – and thereby lost its value as high-tech gadget. This makes clear that the definitions of technology are not fixed but are underlain by changes, negotiations and revaluations. The feminist technology researcher Judy Wajcman (2004) states that every new technology is related to the constitution of gender relations and gender-relevant developments. With every new technology, social power relations and thus gender relations are negotiated.

FabLabs have been discussed as another opportunity to change power relations. Among other things, they are expected to erode the relationship between production and consumption, to establish new relations of production, to change the distribution of power over technology, and so forth. Next, I will analyze the following questions: To what extents are FabLabs gendered as newly constituted spaces? Which gender-specific phenomena can be identified in them? And which prospects for a shift in gender and technology relations exist?

2. New Technologies: Connected to Hopes and Fears

The negotiability and interpretability of 'technology' becomes obvious considering the fact that new technologies are often linked with diverse and even contrary expectations. Recently, such hopes and fears could be observed when the Internet became more popular. From gender and feminist perspectives, three very different prognoses were discussed (Carstensen 2009). One side of the discussion called attention to the Internet as a male domain (Dorer 1997; Spender 1996). This perception of the Internet was influenced by the interpretation of the Internet as a 'technical' application. The central reasons that were cited had to do with the close link between technology and masculinity, the delayed access of women to the Internet, androcentric content (Funken & Winker 2002), and male-dominated discussions in forums and chats (Herring 1996). The Internet was considered to be riddled with the same inequalities and power relations as the 'real world'. At the same time, the Internet was also linked to hopes and expectations for creating solidarity and closer connections between women, including their broader participation in political discussions and decisions. Plant (1997) retold the story of technology and gender, interpreting the Net as feminine, because it refers to female-connoted activities such as spinning, weaving, networking and communicating. Feminists discussed the possibilities of new public spaces and anticipated changes through the removal of the boundaries between the public and private spheres (Consalvo & Paasonen 2002). In addition, worldwide access to information and easier communication were recognized as having the potential to strengthen feminist politics (Floyd et al. 2002; Harcourt 1999). From this point of view, the Internet was regarded rather as a medium than as technology. Furthermore, feminists who were inspired by poststructuralist theories developed utopian projects for a world beyond binary gender relations. Cyberfeminists hoped that the Internet would break down the boundaries between technology and human beings, as well as between men and women. Visions like Donna Haraway's "cyborg" (Haraway 1991) encouraged people to imagine a world without gender. The possibility of anonymous communication via the Internet and 'gender swapping' in chats and forums, where the 'real' body is not present and identities could apparently be invented anew, made the Internet a projection screen for postmodern and deconstructive future designs in which gender relations would be in a state of flux (Bruckman 1993; Turkle 1996).

3. Gendered Aspects of FabLabs

FabLabs are far from being as overloaded with gender-related hopes and fears, as the Internet was when its widespread adoption began. The three expectations that have been outlined nevertheless make it clear in which directions gender and technology relations could shift: towards a male domain, a feminine space or a space for deconstructing gender.

FabLabs are – unlike the Web – 'real' and material spaces where people meet face-to-face and produce things together. Usually, they offer a specific technological infrastructure for digital fabrication, such as a 3D printer and a laser cutter as well as computers. Therefore, they enable invention and production by providing access to tools for digital fabrication, with the idea being "to make (almost) anything" (Gershenfeld 2005, p. ix). FabLabs are places that encourage people to learn collaboratively and to share knowledge and projects – not only in a FabLab but also on the Web with the whole world, similar to the idea of Open Source projects.

Not only do FabLabs suggest that anybody can make (almost) anything, but they are also connected to hopes of democratizing the production of things and returning control over technology production to the users. For Neil Gershenfeld, it is a step towards correcting a historic erroneous trend: the division of society into a few producers and many consumers as well as the division into a few producing countries and many consuming countries (Boeing 2008, 2010). The concept of FabLabs is therefore also a concept of socio-economic development. It is intended to satisfy local needs, to educate children about technology and to empower individuals to create smart devices for themselves, which may not be practical or economical enough for mass production. FabLabs might increase the quality of life in the areas that surround them, so is the hope.

Thus, the idea of FabLabs is explicitly connected to hopes for social, political and economic changes. Ideas of participation, empowerment and democratization play an important role. But how are they connected to gender aspects? Are FabLabs male-dominated rooms for nerds or ambitious engineers? Who are the users? Who has access and who does not? Do the produced artifacts contain gender stereotypes? These questions are discussed below on the basis of existing literature, media reports, research results, as well as on our own empirical findings during the Web-based Work sub-project within the framework of the SKUDI (German abbreviation for Subject Formations and Digital Culture) research project[1]. For SKUDI, we conducted interviews with the operators of the FabLab Fabulous St. Pauli in the St. Pauli district of Hamburg, Germany, and we analyzed the Web presence of twelve FabLabs from different countries. Here I analyze the gendering of FabLabs according to the following categories: access, users, technology, products, education, community and empowerment.

3.1 Access: Between Openness and Exclusion

In terms of gender and technology relations, the question of access to technology and technological spaces is central. Gender and Technology Studies have shown the exclusionary mechanisms of technological cultures. By contrast, FabLabs explicitly claim to be open – but are they open to everyone?

Despite this noble claim, it becomes obvious that the rooms have several access barriers. The interviewees of Fabulous St. Pauli highlight that their door is open during the opening hours. They do not consider access to be a problem, although they are sensitive to the fact that not all demographic groups are represented among visitors to their lab. Estimations differ on how the visitors are distributed according to gender. Some have the impression that there are more male visitors; some say the distribution is nearly equal and the women who come into the FabLab usually come with more concrete ideas and concepts of what they would like to do. Most of the visitors work in creative and artistic professions, which is a more gender-neutral field than technology. Presumably, most of the visitors are academics. Furthermore, the interviewees mention the problem that their FabLab might appear to be too 'nerdy'. One member stated that it is important to him that the FabLab looks quite tidy and does not appear to be a 'half-open garage'.

In general, access is particularly limited by the geographic location, the opening hours, (sometimes) the fees, and culture. Analyses of the Web presences of different FabLabs show a heterogeneous picture. The opening hours as well as the fees differ significantly. As a consequence, in some cases people with low incomes are excluded; in other cases, the limited opening hours prevent people with special working hours or childcare responsibilities from coming (e.g., on a Friday evening). The institutional context of the individual FabLabs, as well as how they present themselves and how they address target groups, seems to be especially relevant

---

1 | SKUDI is a research project funded by Volkswagen Foundation (Germany) and FWF (Austria). Beside our sub project Web-based Work at the Hamburg University of Technology (Jana Ballenthien, Gabriele Winker, Tanja Carstensen), the following universities were involved: University of Klagenfurt (Christina Schachtner), University of Bremen (Heidi Schelhowe) and University of Muenster (Raphael Beer).

Figure 2: Fabulous St. Pauli (Source: Photography by Axel Sylvester).

to the question of access. Some FabLabs are located at universities or research institutes, while others are found in creative, artistic and design milieus, and some are very professional and business-orientated. Some see themselves in the tradition of open workshops, some as part of urban movements that locate themselves in a certain quarter or area and address a particular neighborhood. Some FabLabs see themselves as educational projects, others focus on the development of the area that surrounds them, and yet others emphasize their critique of capitalism. The ways in which they present themselves range from high-tech (MIT) to handicraft spaces for families and children (Munich; Vienna), room for school dropouts (Vigyan Ashram), neighbors and local producers (Hamburg) or professionals (Amsterdam); or they insist that they are not only open to academics (Aachen). In Aachen, for example, women and girls are explicitly addressed, such as when they hold 'Girls' Days'[2]. It is obvious that these different directions address different people and exclude or discourage others.

Some FabLabs therefore experiment with different surroundings and contexts. The most prominent examples are the Mobile FabLab (MIT) or the FabLab Truck (Netherlands). Another example is Fabulous St. Pauli's idea to travel to different places. This idea came about because this FabLab had had no fixed location for an extended period of time. For this reason, and also to address other target groups, they held different 'fab events' at different places – such as at a 'Bauspielplatz' or 'construction playground', where local girls and boys could test the 3D printer. With the playground, they definitely chose a non-male-defined space, as well as a space associated with fun and play.

Gender-specific exclusions are most likely when FabLabs are presented as high-tech factories. Presumably, the most important exclusions do not run along gender lines but rather along education and income.

3.2 Users: Breaking Male Dominance

In terms of Web presences of the different FabLabs, it is noteworthy that not only men are represented in the featured photos. The names of women can also be found among the lists of those who are responsible for the FabLabs. At present it seems as if men are in the majority; however, they are not as overrepresented

2 | These are special days in Germany on which girls do a one-day internship in a male-dominated field in order to motivate girls to work in such professions in the future.

as they are in other technological communities (e.g., hackerspaces). Furthermore, in comparison to F/LOSS (free/libre Open Source software) communities, where women have been a marginalized minority group for many years, men and women are surprisingly often equally represented in Open Hardware/making movements and in Open Fabrication. The OKFestival 2012 addressed this issue in its Gender, Hardware and Open Fabrication session and emphasized that women run some of the most well known Open Hardware/making initiatives. For instance, MIT engineer Limor 'Ladyada' Fried founded Adafruit and now leads it together with Becky Stern from Make: Magazine. Their goal is "to create the best place online for learning electronics and making the best designed products for makers of all ages and skill levels" (Adafruit 2012). Adafruit sells diverse Open Source products, including Arduino packs as well as USB and Android development kits. LittleBits, led by Ayah Bdeir, is an Open Source library of electronic modules that snap together with tiny magnets for prototyping and play (LittleBits 2012). Catarina Mota for example leads the Open Hardware Summit. Mota is also the co-founder of openmaterials.org, a collaborative project dedicated to DIY experimentation with smart materials, and AltLab, Lisbon's hacker space. Her aim is to "encourag[e] people with little to no science background to take a proactive interest in science, technology and knowledge-sharing" (Mota 2012). Alicia Gibb has worked within the Open Source hardware community for the past three years and she is currently in the process of founding and organizing the Open Source Hardware Association (OKFestival 2012)[3].

Many women are also involved in spreading the idea of digital fabrication, hacking and Open Hardware, especially in France. They include journalists such as Sabine Blanc or Ophélie Noor from OWNI[4], for instance, or educators such as the women from nod-A. In Lafayetteville, New York, the first library to become a FabLab grew out of a female initiative. The Open Solar Circuits team is composed of women, too. Specific outreach experiences such as the Mobile FabLab are also female initiatives, among many others. However, the explanation for the differences in participation and access between Open Software and Open Hardware/making movements (OKFestival 2012) is still an open question. And focusing on the engagement of women in Open Hardware projects neglects the participation of people who do not conform to traditional gender distinctions. The question of queer involvement in FabLabs thus remains unanswered.

It is worth noting in this list as well as in the interviews from Fabulous St. Pauli that all those who are engaged in a FabLab have a professional association with digital fabrication. Although the work in the FabLab is mostly voluntary and unpaid (particularly in Hamburg but also in other FabLabs), the jobs of most people are strongly tied to Computer Sciences, the media, the Internet and so forth. As a consequence, this means that the women who are involved can rather be considered pioneers than is the norm in their fields. With this, they might change the culture of the technological field of open and digital fabrication.

3 | For further information see the following websites, viewed 20. September 2012, <http://aliciagibb.com/>; <http://www.oshwa.org/>.

4 | OWNI is a European social media organization that reports on issues in digital society.

3.3 Technology: Towards a Democratic Understanding

The interviews make clear that there are different motivations for taking part in a FabLab and producing things on one's own. One central theme is the fascination for technology. All interviewees have a distinct interest in technology, a passion for problem solving, tinkering and experimenting. New gadgets, especially the 3D printer, fascinate them and they love working with advanced technology, the development of which is currently open. Some have a more philosophical or political interest in technology and are concerned with questions about who controls technology and how technology is involved in power relations.

On the one hand, it becomes clear that FabLabs are strongly perceived as technological spaces. This holds the danger that they could once again be defined and interpreted as male spaces, particularly because they deal with new and advanced technology. Many Web presences also show high-tech products. On the other hand, the interviews as well as the Web presences show that one important idea is to establish a new understanding of technology. Some call it 'more democratic', whereas others say they would like to disenchant or demystify technology and see this as their educational objective (e.g., Munich).

Additionally, a second important motivation is the idea of DIY and producing things on one's own, following the idea "to make (almost) anything" (Gershenfeld 2005, p. ix), to acquire the necessary knowledge and consider what one really needs as an individual. Along with this idea of using and producing technology, hegemonic images of technological progress and a delusional sense of feasibility are undermined. This creates space for new interpretations and definitions of technology and new connotations of technology in terms of gender as a consequence.

3.4 Products: Diverse and Customized

Figure 3: 3D printed objects (Source: Photography by Axel Sylvester).

As mentioned above, Gender and Technology Studies also have to focus on the way gender is inscribed into the artifacts during construction processes. Usually, engineers and computer scientists produce technological objects. They often inscribe their unconscious ideas of gender roles into the artifacts. Therefore, the question is what kinds of artifacts are produced in FabLabs. Looking at the Web presences and the featured artifacts that have been produced, one finds a diversity that cannot easily be presented in an overview. The dominant impression is that they are singular and original – sometimes funny and often focused on individual needs. With examples like wearable computing, which is the design of intelligent clothes, it becomes clear that not only 'nerdy' things are produced.

Since producer and consumer are the same person, consumers use things that contain only their own ideas and not those of any engineer or designer; hence, gender stereotypes are not spread through FabLabs. Moreover, artifacts are individualized and thereby diversified.

3.5 Education: Motivating Girls to Learn about Technology

Increasingly, FabLab experiences are used in educational contexts that address girls' interests and their motivation to learn about technology. Dlodlo and Beyers (2009) focus on the issue of inequality in girls' and women's access to education and careers in science, engineering and technology (SET). Using the example of secondary schoolgirls in South Africa, they investigate how their SET awareness changed after they were exposed to a hands-on rapid-prototyping environment in a FabLab. The results show that the girls' confidence in being able to handle technology-related tasks was enhanced: "The girls indicated that this was an interesting session that stimulated their interest in computers and technology. They indicated that they were keen to learn more about computers as a result of this stimulating exposure. They were excited by the sense of achievement in their ability to design a product from scratch and produce a high quality prototype in such a short period of time. This brought them a degree of satisfaction" (Dlodlo & Beyers 2009, p. 429).

Similarly, the Marymount School of New York (a college preparatory Catholic day school for girls) implemented a FabLab with the aim of closing the gender gap in careers in SET and mathematics: "We aim to cultivate a problem-solving, collaboration, innovation, and entrepreneurship mindset. We want our students to be doers and makers" (Hasselle 2012).

The FabLab Aachen, Germany held 'Girls' Days' in 2010 and 2011. Their aim was to show girls that computer science is not limited to software and computers, but can, for example, also include wearable computing. At the workshop, the participants produced T-shirts with flashing patterns. It is yet to be seen how successful these efforts to motivate girls about technology will be in the long run. However, FabLabs may have the potential to offer more convincing technological experiences than previous projects did.

3.6 Community: Bodily Interactions

Whereas thoughts about the Internet focused on the hope that bodiless interaction might encourage identity experiments, as well as on hopes of new gender roles and decreasing gender relevance, FabLabs, by contrast, are 'real' spaces, which of course awaken other hopes. The interviewees emphasize that it is important that this is a kind of 'real' room where participants actually meet other people and work together collaboratively. Therefore, it is important to them to come together at the same time and at the same place to share knowledge and to solve problems collectively – instead of sitting alone at home with their own 3D printer.

This implies that visible bodily attributions like gender, race, age, and ability cannot be hidden or ignored. They always play a role in face-to-face interactions and will thus also have effects on the interaction in FabLabs. At the same time, however, the hopes that surrounded the Internet were not fulfilled. Early studies showed that gender was in fact one of the most important categories in 'virtual' communication (Döring 2008).

3.7 Empowerment, Participation and Political Perspectives

Not least, the idea of FabLabs is a political one. It encompasses critiques of the capitalist relations of production, the distribution of power over technology, of excessive affluence, of the global division of labor, and so on. It dovetails with political visions of commons, of urban movements, of local production and of shifting the division between production and consumption. FabLabs have a clear mission to empower people to participate and engage in technological development and to educate underprivileged people. For the interviewees of Fabulous St. Pauli, the political dimension of fabbing is a very important motivation. They are concerned with social and political problems such as social inequality, abundance and gentrification. (Read more about St. Pauli's FabLab in the article of Sylvester and Döring in this book.) Through their engagement in the FabLab, they fight for democratic education, democratic technology and, like Fritjhof Bergmann (2005), they see opportunities for an alternative economy.

All of these issues have rarely been linked to feminist ideas and projects. However, it is clear that ideas of empowerment, participation and critique of the existing division of labor have a lot of relevance to feminist claims. It might be promising to link both ideas more closely.

Figure 4: Writing with 'Stanzgerät Ideal No. 1' (Source: Photography by Axel Sylvester).

4. Conclusion

Although the relationship between gender and technology remains stable in many respects, studying FabLabs allows for some optimism. There are some points that suggest that FabLabs are male-dominated spaces – because they deal with advanced technology, they might be considered nerdy in the conventional male-orientated sense, and they might have inappropriate opening hours for people with childcare responsibilities – yet a number of the aspects under consideration suggest that FabLabs might also become spaces for women, or in the best case spaces where gender plays no significant role. Judging by the attempt to establish a new, more democratic understanding of technology and also judging by the large number of women who are already engaged as well as by their proximity to feminist ideas, one can hope that FabLabs will become comfortable spaces for women and girls and hopefully for all genders. Due to their aspirations towards education and projects for girls, they actively encourage a change in the relationship between gender and technology.

It seems as if an opportunity to negotiate gender relations in a field of new technology is currently emerging. If that is the case, shifts might be possible in FabLabs. In the worst-case scenario, FabLabs will not become deemed male domains because they are not considered to be promising, high valued and profitable. More optimistically, they are spaces to develop and negotiate some revaluations among gender and technology beyond the 'old' connection between technology, masculinity and power. These chances should be considered more consciously in the future so as not to lose the clear potential that FabLabs currently have. Hence, questions of gender should not be limited to considering men and women only; instead, FabLabs should also be considered possible queer spaces. Furthermore, the idea of negotiating power relations and issues of inequality in the context of FabLabs should not be limited to gender issues. Disability is an issue that has already been addressed by FabLabs. The analysis has shown that other power relations might be equally important concerning the access and use of FabLabs, particularly people's educational background and income. It is widely unknown how open and comfortable FabLabs in Europe and North America are for people of color. Urban/rural differences might also be relevant. An awareness of these different power and inequality relationships would offer an opportunity to connect the political ideas of the FabLab movement with other movements and to establish a new relationship between technology and power.

Acknowledgement

I would like to thank Jana Ballenthien, Anna Köster-Eiserfunke, Bertold Scharf and Gabriele Winker, with whom I worked on the FabLab research in the context of SKUDI. I would also like to thank the Volkswagen Foundation for funding SKUDI. And lastly, I would like to thank all the interviewees from Fabulous St. Pauli for donating their time and thoughts to our research.

References

Adafruit 2012, viewed 31 September 2012, <http://www.adafruit.com/about/>.

Berg, A-J & Lie, M 1993, 'Feminism and constructivism: Do artifacts have gender?', *Science, Technology and Human Values*, vol. 20, no. 3, pp. 332-351.

Bergmann, F 2005, *Neue Arbeit – Neue Kultur*, Arbor, Freiburg, Germany.

Boeing, N 2008, 'Die Marx-Maschine',viewed 31 September 2012, <http://www.freitag.de/autoren/der-freitag/die-marx-maschine?p=1&-C=>.

Boeing, N 2010, 'The future is fab', *Technology Review*, viewed 31 September 2012, <http://www.heise.de/tr/blog/artikel/The-future-is-fab-942575.html>.

Bruckman, A 1993, 'Gender Swapping on the internet', *Proc.INET'93*, pp. EFC1-EFC5.

Carstensen, T 2009, 'Gender trouble in web 2.0: Gender relations in social network sites, wikis and weblogs', *International Journal of Gender, Science and Technology*, vol. 1, no. 1, <http://genderandset.open.ac.uk/index.php/genderandset/article/view/18>.

Cockburn, C 1986, *Machinery of dominance. Women, men, and technical know-how*, Pluto Press, London, UK.

Cockburn, C & Ormrod, S 1993, *Gender and Technology in the Making*, Sage, London, UK.

Consalvo, M & Paasonen, S (eds.) 2002, *Women & everyday uses of the internet. Agency & identity*, Peter Lang, New York et al..

Dlodlo, N & Beyers, RN 2009, 'The experiences of South-African high-school girls in a Fab Lab environment', *Proceedings of World Academy of Science, Engineering and Technology*, vol. 25, pp. 423-430.

Dorer, J 1997, 'Gendered Net: Ein Forschungsüberblick über den geschlechtsspezifischen Umgang mit neuen Kommunikationstechnologien', *Rundfunk und Fernsehen*, vol. 45, no. 1, pp. 19-29.

Döring, N 2008, 'Männlichkeit und Weiblichkeit im Netz: Dimensionen des Cyber-Gendering', in von Gross, F, Marotzki, W & Sander, U (eds.), *Internet – Bildung – Gemeinschaft*, VS Verlag für Sozialwissenschaften, Wiesbaden, Germany, pp. 119-141.

Floyd, C, Kelkar, G, Klein-Franke, S, Kramarae, C & Limpangog, C (eds.) 2002, *Feminist challenges in the Information age. Information as a social ressource*, Leske & Budrich, Opladen, Germany.

Funken, C & Winker, G 2002, 'Online-Aktivitäten von und für Frauen im deutschsprachigen Internet', in Bundesministerium für Familie, Senioren, Frauen und Jugend (ed.), *WOW – Women on the Web. Internationale Konferenz 8.-10. März 2001*, Berlin, Germany, pp. 25-38.

Gershenfeld, N 2005, *Fab. The coming revolution on your desktop – from personal computers to personal fabrication*, Basic Books, New York.

Haraway, D 1991, *Simians, cyborgs, and women. The reinvention of nature.* Routledge, New York.

Harcourt, W 1999, *Women@internet. Creating new cultures in cyberspace*, ZedBooks, New York.

Hasselle, D 2012, 'All-girls school designs way to close gender gap in math and science', *DNAinfo New York*, viewed 31 September 2012, < http://www.dnainfo.com/new-york/20120521/carnegie-hill/all-girls-school-designs-way-close-gender-gap-math-science#ixzz26whHguLB>.

Herring, SC 1996, 'Bringing familiar baggage to the new frontier: Gender differences in computer-mediated communication', in Vitanza, VJ (ed.), *CyberReader*, Allyn & Bacon, Boston, pp. 144-154.

Knoll, B & Ratzer, B 2010, *Gender Studies in den Ingenieurwissenschaften*, Facultas, Vienna, Austria.

LittleBits 2012, viewed 31 September 2012, <http://littlebits.cc/>.

Mota, C 2012, viewed 31 September 2012, <http://summit.oshwa.org/about-ohs/cm_sq/>.

OKFestival 2012, *Gender, hardware and open fabrication*, <http://okfestival.org/gender-hardware-open-fabrication>.

Oudshoorn, N, Saetnan, AR & Lie, M 2002, 'On gender and things: Reflexions on an exhibition on gendered artefacts', *Women's Studies International Forum*, vol. 25, no. 4, pp. 471-483.

Paulitz, T 2012, *Mann und Maschine. Eine genealogische Wissenssoziologie des Ingenieurs und der modernen Technikwissenschaften, 1850-1930*, transcript, Bielefeld, Germany.

Plant, S 1997, *Zeroes and ones: digital women and the new technoculture*, Doubleday, New York.

Spender, D 1996, *Nattering on the net*, Garamond Press, Toronto, Canada.

Turkle, S 1995, *Life on the screen. Identity in the age of the internet*, Simon & Schuster, New York.

van Oost, E 1995, "Male' and 'female' things', in Brouns, M, Verloo, M & Grunell, M (eds.), *Women's Studies in the 1990s. An introduction to the different disciplines*, Couthinho, Bussum, Netherlands, pp. 287-310.

Wajcman, J 1991, *Feminism confronts technology*, Polity Press, Cambridge, UK.

Wajcman, J 2004, *TechnoFeminism*, Polity Press, Cambridge, UK.

Zachmann, K 2004, *Mobilisierung der Frauen. Technik, Geschlecht und Kalter Krieg in der DDR*, Campus, Frankfurt/Main, Germany.

Zorn, I, Maaß, S, Rommes, E, Schirmer, C & Schelhowe, H (eds.) 2007, *Gender designs IT. Construction and deconstruction of information society technology*, VS Verlag für Sozialwissenschaften, Wiesbaden, Germany.

Figure 1: Workshop participants presenting the results of the 2D design session. Motives designed on the computer were cut out on the vinyl cutter and pressed on the T-shirt with a thermo transfer press (Source: Photography by Happylab).

FABRICATING ENVIRONMENTS FOR CHILDREN

IRENE POSCH

1. Introduction to Digital Fabrication

One of the main goals of FabLabs is to be an open space providing access to modern tools for invention (Gershenfeld 2005, 2012). Instruments beyond the affordability of individuals are publicly made available, on hand for hobbyists, students, professionals, and artists – everyone with the objective to experiment on products and procedures with digital fabrication technologies. Policies of what open access means and whom it does include greatly vary based on the institutions hosting the FabLab and the organizational model they deploy. The majority of FabLabs are targeted at students or adults working on individual or collective projects, often excluding younger children from working independently in the lab. Some of the tools and machines are potentially dangerous for the children themselves as well as for other people present, if they are not operated correctly. For security and liability reasons, institutions therefore do not accept children without a supervisor or trained mentor.

At the same time, there are many reasons why a FabLab could be of special interest for children. As digital fabrication technologies are becoming more and more important (Mota 2011), ongoing discussions frame how to publicly make them accessible in formal and informal educational settings like libraries, museums, and schools (Terrone 2011; Posch et al. 2010; Eisenberg 2011). While educating and learning from each other is an important aspect of the FabLab network, working with young users at a FabLab is a comparably new field. Given the possibilities and capabilities of digital fabrication technologies, they seem to be very promising for hands-on learning approaches in STEM-related disciplines as well as for design and artistically motivated creations.

This inclusion of emerging fabrication technologies in learning environments and the important role they could play in science and engineering education is increasingly being discussed (Eisenberg & Buechley 2008; Dlodo & Beyers 2009). Paulo Blikstein will introduce the potential of FabLabs in schools and their integration into the curriculum later in this book. I will focus on what can be done in a publicly accessible community space, establishing the FabLab as a creative and safe informal learning environment, also welcoming children apart from its adult users.

Not being bound to external educational objectives, we designed workshops covering essential aspects of FabLabs, such as 2D and 3D design and fabrication, electronics and programming as out-of-school activities, engaging children in hands-on activities. We want to provide children with a setting where they can use digital fabrication technologies for individually motivated construction processes in order to make participants more self-confident in operating them. We also hope to encourage them to think and explore potentials of digital fabrication technologies in their own lives (Katterfeldt et al. 2009; Blikstein 2008).

The selection of software and hardware used in the workshops aims to provide insight into a lab's possibilities. At the same time, we want the experiences to go far beyond the FabLab. This implies a selection of free and platform-independent software environments to increase the transferability of learned skills to computers anywhere at home or in schools. A second important base is the production of artifacts participants can take with them at the end of the workshop regarding the production time, output dimensions as well as costs involved.

1.1 Context, Content and Creation

A crucial element of the workshop consists of providing contextual information about current and potential usage of digital fabrication technologies. This includes the functionality of machines, haptic examples of objects previously produced in the lab and graphical material giving an insight into what has been done beyond the borders of the lab. Being a lab that is used by adults and children alike, it also gives an insight into its potential beyond its use as educational infrastructure. Most importantly though, the introduction opens a space for discussion where children can articulate their thoughts and expectations how digital fabrication could be useful in their personal context.

Among the most important achievements of digital fabrication technologies is its implementation of objects of personal need that are not covered or beyond the reach of the mass market. Therefore, we want to preserve this important aspect, giving children the possibility to make something of personal use they might not get elsewhere. Within the scope of the workshop modules, they work on individual projects rather than repeating predefined tasks.

A FabLab is a place to make almost anything, and we encourage children to make as much as possible themselves – not only generating ideas but also designing adequate data and operating the machines. The goal is to show potentials and difficulties in dealing with proposed technologies. Being able to master them, with guided help where necessary is a fulfilling experience, while it also gives a realistic insight into skills needed in working with the machines and getting to know their limits.

2. Children's Practice in the FabLab

Based on the principles explained above, we run two main programs: an introduction workshop spanning two days with directed modules giving an overview of the lab's possibilities, and a semester-long program allowing children to come to the FabLab once a week, work on their projects, and get help and guidance where needed. The target groups are children between ten and 15 years old; the youngest just finished elementary school, the older ones just are about to decide what education or professional career to follow. One or two mentors, depending on the amount of participants, are present to guide the group of up to 12 students. The deliberate mix of ages is meant to encourage older and younger students to interact and help each other. The workshops take place in the so-called 'Happylab' Vienna FabLab, Austria.

Figure 2: Becoming of a project, realized in the KidsClub. A participant decided to make a tablet stand as a birthday present for his grandfather. First, a prototype is cut out of cardboard with the laser cutter to test the design. After minor adjustments, the final product is produced in the material of choice, in this case, plywood (Source: Photography by Irene Posch).

2.1 Forms of Engagement

Children completely new to the FabLab mostly register for the two-day introduction workshop. In order to give a comprehensive – yet suitable for young users – overview of a FabLab's possibilities, the workshops consist of four individual modules introducing 2D and 3D design and fabrication, PCB (Printed Circuit Board) fabrication and assembly, and software programming. It builds upon a selection of common online software: Gimp and Inkscape to design a motif, cut it out on the vinyl cutter and print it on a T-shirt, SketchUp to digitally design a 3D model and print it out on a 3D printer, Scratch, a programming language targeted at young users (Resnick et al. 2009) to get an insight into programming, and making a printed circuit board and assemble necessary parts for a Drawdio, an electronic circuit translating electric resistance into frequency/sound (Silver 2009).

Figure 3: Using SketchUp to digitally design a 'dream house' (left) that is subsequently printed out in 3D (right) (Source: Photography by Irene Posch).

Figure 4: Signature stamp designed by one of the participants, copying his own handwritten signature (left) and a jewelry set, earrings and pendant, designed and fabricated in the KidsClub (right) (Source: Photography by Irene Posch).

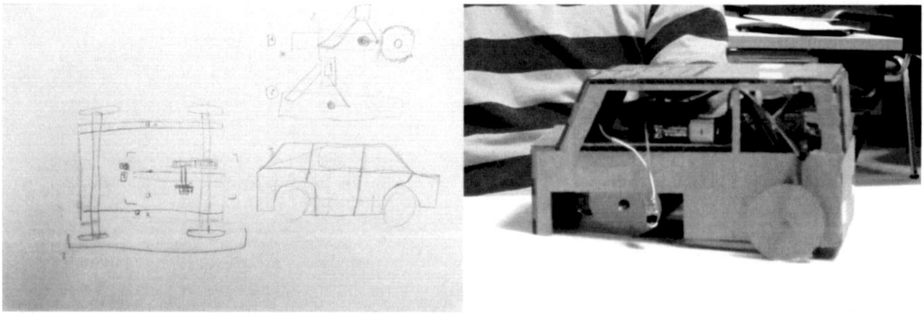

Figure 5: Sketches for an Arduino controlled car (left) and the first prototype of the car (right). The auto body consists of laser-cut cardboard. Basic functionalities for driving forward and turning are already included (Source: Photography by Irene Posch).

As output, they take home their individually designed T-shirt, a printed model of their 3D design and the self-assembled electronic music instrument. The result of the programming module is stored online so that participants can access it anytime and can continue working on it later on if they wish. The modules explain the output and the machines to use, however, it is a crucial part of the workshop that children come up with their own ideas.

Children's reactions towards the introduction workshop are very positive – a lot of them ask when they can come back and make more things, often already having specific projects in mind (Posch & Fitzpatrick 2012). As most of them are still too young to come to the FabLab alone, and most parents do not have the resources or knowledge to accompany their children, we founded the KidsClub, opening the lab for children one afternoon a week. With the help of mentors, we welcome them to work on their individual projects, thus overcoming the limited time and choices available in the introduction workshops.

The KidsClub is targeted at children who want to use the FabLab's infrastructure for the implementation of projects of their own choice. It opens up considerably more space for children's own ideas and allows for more freedom in software and machinery, as well as more time to work on projects. This freedom also turned out to be difficult for some children in finding what they actually want to make and what skills they need to reach their goal. In starting off with smaller projects, they increasingly gained self-confidence in finding solutions as to how to approach bigger projects. Their results vary decisively, they span from decorative accessories to functional parts, from objects of personal use to projects relevant in their family or home context and repairing: children were working on tracking systems for their turtle, personalized door locks or board games as well as digital games for their siblings. A few projects implemented in the workshop programs are pictured in this article. Only a few students could be convinced to also document and share their project on a weblog online (HappylabKidsClub 2012). We hope to be able to improve this in the future in order to establish an exchange of ideas among the children, possibly even beyond the borders of the lab.

2.2 Children's Interests and Understanding

Feedback and observations showed that children were most engaged in making something unique that has a meaning in their everyday life (Posch & Fitzpatrick 2012). The children thought that the T-shirt and electronic projects, as well as the projects they planned and produced in the KidsClub were especially useful and motivating. As main motivation, they mentioned making something themselves and being involved in every step of the projects: from generating the idea to the final object. Their reactions included statements about being 'proud of the result because I made it myself' and satisfied because 'it looks as if it was bought at a shop'. Based on their own artifacts, many children also think about how to extend the knowledge gained. They are discussing what their friends and others might like and what they could eventually sell them. This was especially true for projects including the laser and vinyl cutter and electronics. For many of the participants, the personal use of their 3D models was much less clear, given we just had the ability to print objects of small size made of plastic that could not always stand up to the expectations they had.

Moreover, 3D printing technologies were by far the most discussed technologies regarding their overall future influence and what they would eventually do with them. Children remarked possible influences for consuming culture as follows: 'Will we just sit at home and order things?' and 'Wouldn't it be very bad for the economy, since a lot of small businesses would go bankrupt if people could make everything at home?' and 'What if someone starts to print out bad things, like weapons?' The children were also afraid that machines might replace human skills one day. Some even commented on potential negative environmental effects and waste production and came up with ideas of how to use sustainable materials. On the whole, 3D printing technologies appeared very interesting for reasons of technical feasibility, less so for the actual use children could make of it.

Another aspect of interest is that although not connected to any school curriculum or specific learning outcome, children drew connections to formal education and where technologies present in a FabLab might be of importance and interest in their future. They related their experiences at the FabLab to fields like mathematics, physics, and computer science as well as to arts, crafts, and geometry classes, among others. This could be an interesting path to follow, in order to integrate science- and arts-related education into a holistic learning approach.

3. Conclusion

In conclusion, a FabLab has a lot to offer for children. Focusing on making the lab and its machines accessible and safe for young users, it can be a rich environment for experimentation and associated learning. While it proved necessary to teach basic principles to provide a base to elaborate from, children's ideas carried them on to learn and ask for help about what they need to realize their projects. The diversity of projects and ideas shows that children can quickly adapt and make personal use of the technology proposed.

Children were excited to work in the lab, however, this may be (partially) influenced by the fact that FabLabs are novel to them. FabLabs and their inventory are still beyond the reachability and even knowledge of great parts of the population. Having access to it and being able to implement personal projects is an exclusive privilege, as machines are not widely accessible yet. Offering workshops for children in a public FabLab as described, we hope to increasingly change that, given that not many schools are yet prepared to incorporate necessary technologies in their classes. The diversity of projects coming into being and evolving discussions among the children about the potential of digital fabrication technologies are an especially encouraging factor to continue and extend the work in this field. They are also indicators that hands-on experiences and the possibility for free explorations, as well as critical reflection in an environment which encourages innovation, mark an important step towards active and informed usage of future technologies. Not exclusively but particularly for children.

Acknowledgement

I would like to thank INNOC – Austrian Society for Innovative Computer Sciences, for the initiative of building and operating the Happylab – Vienna FabLab and their support in running youth activities there. Further thanks to the Austrian Federal Ministry of Economy, Family and Youth and the department for culture, City of Vienna for enabling the series of workshops and children programs.

References

Blikstein, P 2008, 'Travels in Troy with Freire: Technology as an agent for emancipation', in Noguera, P & Torres, CA (eds.), *Paulo Freire: the possible dream*, Sense, Rotterdam, pp. 205-244.

Dlodlo, N & Beyers, R 2009, 'The experiences of South- African high-school girls in a FabLab environment', *International Journal of Social and Human Sciences*, vol. 3, pp.1-15.

Eisenberg, M & Buechley, L 2008, 'Pervasive fabrication: Making construction ubiquitous in education', *Journal of Software 2008*/III4, pp. 62-68.

Eisenberg, M 2011, 'Educational fabrication in and out of the classroom', *Proceedings of Society for Information Technology & Teacher Education International Conference*, AACE, Chesapeake, VA, pp. 884-891.

Gershenfeld, N 2005, *Fab. The coming revolution on your desktop – from personal computers to personal fabrication*, Basic Books, New York.

Gershenfeld, N 2012, 'How to make almost anything: The digital fabrication revolution', *Foreign Affairs* 91/2012, pp. 43-57, viewed 22 October 2012, <http://www.foreignaffairs.com/articles/138154/neil-gershenfeld/how-to-make-almost-anything>.

Katterfeldt, E, Dittert, N & Schelhowe, H 2009, 'EduWear: smart textiles as ways of relating computing technology to everyday life', *Proceedings of the 8th International Conference on Interaction Design and Children*, ACM, New York, pp. 9-17.

HappylabKidsClub 2012, *Online blog and documentation accompanying the KidsClub*, viewed 22 October 2012, <http://happylabkidsclub.wordpress.com/>.

Mota, C 2011, 'The rise of personal fabrication', *Proceedings of the 8th ACM conference on Creativity and cognition*, ACM, New York, pp. 279-288.

Posch, I, Ogawa, H, Lindinger, C, Haring, R & Hörtner, H 2010, 'Introducing the FabLab as interactive exhibition space', *Proceedings of the 9th International Conference on Interaction Design and Children*, ACM, New York, pp. 254-257.

Posch, I & Fitzpatrick, G 2012, 'First steps in the FabLab: Experiences engaging children', *Proceedings of the 23rd Australian Computer-Human Interaction Conference*, ACM, New York, pp. 497-500.

Resnick, M, Maloney, J, Monroy-Hernández, A, Rusk, N, Eastmond, E, Brennan, K, Millner, A, Rosenbaum, E, Silver, J, Silverman, B & Kafai, Y 2009, 'Scratch: programming for all', *Commun*, ACM 52, 11/2009, pp. 60-67.

Silver, J 2009, 'Awakening to maker methodology: the metamorphosis of a curious caterpillar', *Proceedings of the 8th International Conference on Interaction Design and Children*, ACM, New York, pp. 242-245.

Terrone, P 2011, *Is it time to rebuild & retool public libraries and make "TechShops"?*, viewed 22 October 2012, <http://blog.makezine.com/2011/03/10/is-it-time-to-rebuild-retool-public-libraries-and-make-techshops/>.

MATERIALITY
AND
VIRTUALITY

NOTES ON MATERIALITY AND VIRTUALITY

BRUCE STERLING

I am Bruce Sterling, an author and journalist. I first heard about FabLabs when the project was started at Massachusetts Institute of Technology (MIT) and found the idea quite ambitious for an academic project. There is something intriguing about the idea of dropping unusual technology packages into social situations where people would have few expectations about them. It is an interesting experiment in material culture, rather like the One Laptop Per Child project, which also had its origins at MIT.

I believe that the common distinction between virtuality and materiality will vanish within the next hundreds of years. Moreover, I think that even in as soon as 30 years from today these concepts will sound quite old-fashioned. In my opinion, FabLabs will not be labs anymore twenty years from now, but they will simply be some everyday part of how things are designed and made and, hopefully, recycled.

Personally, I am mainly concerned with the concept of 'spimes'[1], which has little to do with FabLabs as such. Spimes are a design theory about using ubiquitous computing in the service of sustainability. One of the ideas of spimes is that digital fabrication or 3D printing would become the major form of manufacturing, because it is easier to control the flow of materials that way. The other ideas behind spimes are concepts and design specs, unique identities for objects, the use of fabricators, tracking technologies, searching technologies and recycling, all of these very tightly organized through comprehensive machine surveillance and data-mining. Nobody has ever built a real spime, it's a visionary proposal about a different kind of industrial base, one that is ubiquitous and sustainable.

1 | Spimes are "manufactured objects whose informational support is so overwhelmingly extensive and rich that they are regarded as instantiations of an immaterial system" (Sterling 2005, p.11; explanatory note by the editors). Sterling, B 2005, *Shaping Things*, MIT Press, Cambridge, MA.

Figure 1: Casey Reas 2008 (Source: Courtesy of the artist).

CONSIDERING ALGORITHMICS AND AESTHETICS

FRIEDER NAKE

1. Introduction

Mathematicians have their most creative period in their young twenties, and after the age of fifty they do not come up with new ideas. They mature young, or never. This may be true or not, but the inventive and successful mathematician Godfrey Harold Hardy fell into depressions and tried to commit suicide when he felt the end of his creative lifespan had come. Ambition and passion are the "proper justification of a mathematician's life," he wrote (Hardy 1967, p. 65). Results in mathematics are different from those in other disciplines: once found, they remain true. They may not be important in later years, or they may be shown to be special cases of some more general theorem. However, once proven, and thus created, they persist.

Hardy compares the mathematician to painters and poets. He writes: "A mathematician, like a painter or a poet, is a maker of patterns. If his patterns are more permanent than theirs, it is because they are made with *ideas*. A painter makes patterns with shapes and colors, a poet with words. A painting may embody an 'idea', but the idea is usually commonplace and unimportant. In poetry, ideas count for a good deal more; but ... the importance of ideas in poetry is habitually exaggerated" (Hardy 1967, p. 84; his emphasis).

The patterns these three makers make all are, in some way or another, connected to ideas. These ideas, Hardy submits, are not really very important in the case of painting. In the case of poetry, they are exaggerated. But in mathematics, they are the very form of the patterns themselves. The idea and its pattern become one and the same, he suggests, and many of us may find this intriguing. We must admit, however, that even in mathematics, the idea must take on *some* material form in order to be perceivable, however weak its materiality.

Coarse as such analyses always are, we may read Hardy the following way. Humans love making patterns. Their creative capacities allow them to come up with quite different sorts of patterns. Patterns are expressions of ideas. One way to distinguish them is the distance of the pattern's form from the idea appearing in the pattern. Hardy suggests a scale from logical (near zero distance) via poetic (more) to painterly (much). We may paraphrase this as: formula, text, object. In FabLabs all of these appear to be present and play important roles: formulae are needed to do the programming; text becomes the narrative around the object of design; material object is the final goal of fabrication.[1]

If we interpret the zero-dimensional pixel as the final result of abstraction and as the appearance of pure thought in the un-sensuous form of computability, then the algorithmic drilling, moulding, constructing, kneading, sculpting, or milling of artificial and amorphous materials into space-filling shapes (the FabLab's occupation) stands out as an embodiment of pure thought. But FabLab is an everyday approach to cultural production, mainly of trivia. Instead of *return-to-sender* a *return-to-matter*.

1 | There appears to be a parallel to Flusser's anthropology of cultural techniques: four dimensions of the space of our bodily lived experience, three dimensions of architecture and sculpture, two of the picture, one is linear text, and zero dimensions is the world of pixels (Flusser 1983).

This essay offers a theoretical perspective on artistic production supported by computer technology, garnished with a remark on the history of algorithmic art (in sections 2 and 3), and a selection of mathematical things or techniques, adorned by a bit of aesthetics. I hope the reader enjoys, but unfortunately I cannot guarantee it.

2. Ontological State of Art and Technology

"The aim of generative aesthetics is the artificial production of probabilities of innovation or deviation from a norm." (Bense 1971, p. 57) The German philosopher and founder of information aesthetics, Max Bense (1971), first presented this thought in public on the 5th of February, 1965, at the opening of the first exhibition of computer art.[2] The mathematician Georg Nees was exhibiting a selection of small-size drawings which were made with computer programs. In those early days of the algorithmic revolution (Peter Weibel's term), the computer's output was a punched paper tape that was fed off-line into an automatic drawing machine. Nees used a Zuse Z64 Graphomat for the purpose (also called 'flat-bed plotter').

That first exhibition took place in the premises of Bense's Aesthetisches Kolloquium at the University of Stuttgart (then still an Institute of Technology). Two exhibitions of computer art followed the same year in New York (A. Michael Noll) and Stuttgart (Frieder Nake).

Remarkable for our context is Bense's inauguration of the term 'generative aesthetics'. He envisioned a time when algorithmic thinking and creating would become a common activity, perhaps for many. Manual components of the creative process would be carried out by properly programmed computers. Mathematical foundations would remain a human task.

But Bense left room for speculation about where the line between the human mind and the mechanic processor had to be drawn. His aim of generative aesthetics is "the *artificial* production" (Bense 1971; my emphasis) of probabilities. Since it is probabilities that get produced, we are not talking about the eventually visual, and therefore material, pieces that are going to be presented to the public. Bense's probabilities are a decisive ingredient needed in the generative process *before* it enters its material phase. Bense's formulation is aiming at the probability distributions needed to control the generation of sequences of pseudo-random numbers that are essential for almost all of generative art.

Bense is thinking of the *artificial* production of those probability distributions, so he regards them as the result of an automatic process by the computer. Incidentally, he uses the term 'artificial' a second time at Nees' opening. This was when he coined the term *artificial art*. He did so to calm artists' mood when they revolted against seeing the computer as a creative partner.

However, the interpretation of that important formulation at the birth of computer art is not open. Even if we write algorithms that automatically generate

2 | The original publication is in German in a small brochure, inaccessible today. It is included in Bense's main work on asthetics (1965). For the Cybernetic Serendipity show in London, 1968, it first appeared in English.

probability distributions, those algorithms would become our human contribution. Even if we consider human beings to have nothing to do with the final material production of the work, human beings will always remain the last to make productive decisions. It is absurd to even speculate about where to 'find' creativity.[3]

Both technology and art are human-made products. As artifacts they share some properties and are different than others. They share the ontological status of reality which here is the reality of the artificial. The artificially *made* thing shares with the naturally *given* thing the status of reality. Real is what is as it is, and nothing else. But the artificial may ontologically be further recognized as *co-reality*, as Bense suggested (1965). Co-reality is that kind of reality that *comes with* reality. It presupposes reality as its status in the sense just mentioned: that that is so and not different. But in its so-being, it transcends exactly this status by being more. The car is a pile of tin, iron, instruments, engine, screws, bolts, plastic, leather, rubber, and much more in an intricate and purposeful structure. Purpose and structure is what comes *with* the material.

This also applies to paintings in a similar way. It is canvas, oil, wood, or plastic in a purposeful arrangement that makes it a painting. The material analysis is important to find out when the painting was made. But what it was made for and why, you will not find by measuring, instead, you must interpret things – 'co-reality' fits well in both cases. However, Bense (1965) introduces another distinction: Technology is *necessary* co-reality, art is *accidental* co-reality.

3. Virtuality Supporting Actuality

There was a time when surprisingly many intelligent people claimed that virtual reality was a phenomenon *opposed* to reality. Certain results of programming were called *virtual reality* for the only reason that they appeared as still or moving images depicting three-dimensional scenes and living creatures. A mad discussion emerged around this phenomenon. Origin and results of the debate were rather simple.

Reality, as we had known it, now appeared as split into two modes: actuality and virtuality. *Actuality* is that mode of reality which unfolds into all aspects of our environment. It is reality in full flesh, for all our senses to perceive. *Virtuality* is that mode of reality that is in one or several aspects restricted. At least one, if not more of our senses are not and cannot be involved in perceiving that mode of reality.

Something can be real but can be expanded to reach full-fledged actuality status from virtual mode. And something can be real, but can be reduced to lose some of its sensuality and, thereby, fall back into virtual mode.

Once the seeming antagonism of reality vs. virtual reality had been dissolved into reality as the unity of actuality and virtuality, it was trivial to realize that we had always already been surrounded by great forms of virtuality. The poem, the drama and the novel, the film, the opera, the childrens' play, as well as the songs and symphonies, the drawing, the painting, the sculpture now all appeared as forms of virtuality.

3 | For two recent books, see Boden (2010) and McCormack and d'Inverno (2012).

The particular reductions of actuality demonstrated in these forms lead to possible insights, interpretations, and emotions. Religion and other utopia appeared as versions of virtuality.

The world of software seemed to encompass many, if not all, of those virtualities. Anything on the computer was, in some sense, a duplication of a world phenomenon (into actual and virtual) (Nake 1984). Software revealed its character as algorithmic sign (Nake 2001, 2009). As of right now, things owe their existence to a running computer; another way of expressing this is to realize that they possess *surface* and *subface*. The surface is perceivable by us with one or several of our senses. The subface is computable by the computer making use of processor and storage capacities.

In the digital realm of reality nothing exists without the split into these two, i.e. as their unity. The virtual mode of reality (accessible by the machine) is the pre-condition for the actual mode of reality (accessible by us) to appear. Virtuality supports actuality by transforming the world into algorithmic form. Concrete actuality disappeared in the abstract virtual world of software. Now it is re-appearing as a derivative of virtuality.

Computer art became the first form to explore this new mode of reality. Other forms followed during the gradual development of digital media, including video games. The three-dimensional world of computer milling fabrication is a fashionable current trend.

4. Beautiful Mathematics

Mathematics is largely an activity of the mind. Clearly, there is no pure mind and mind does not exist isolated from body. But once philosophers had separated mind from body as two capacities of the human being, mathematics turned out to be the sort of activity that could justifiably be identified as a mental activity of great purity.

The more purely mental an activity is, the less it seems to exist in the aesthetic domain. For aesthetics deals with that that pertains to the senses. In the modern view of aesthetics, as it was introduced by Alexander Baumgarten (1986), aesthetics is the science of sensual cognition. For instance, mathematics cannot be perceived by our senses but only thought by our brains, so there is no aesthetics in mathematics. But people do speak of aesthetic sensation in connection with mathematics, and they talk about the beauty of mathematics. There must, in mathematics, be connections between operations of our brains and sensations of our senses.

In this section, I consider a few emanations of the mathematical mind that do affect the senses and thus may justify talks about 'beautiful mathematics' – a term that is certainly questionable. There will be very few technical descriptions, since this is not a book on mathematics. A bit of formalism is needed nevertheless: it is itself beautiful.

4.1 Fractals

The mathematical phenomenon known as fractal attained an immense popularity in the 1980s. This was due to Benoît Mandelbrot's (1983) new systematic investigation of certain point sets in the plane. Some (like the Cantor set, the von Koch curve, the Peano curve) had been known up to one hundred years before. As mathematical 'monsters' they contradicted intuitive expectations of their behavior. Now that computers were available, Mandelbrot ran experiments involving long chains of seemingly simple iterative calculations that showed dramatic results. He used graphic techniques to depict results of his program's calculations. Such visualizations were taken up by hundreds of researchers and thousands of students.

Figure 2: The Mandelbrot set and five states of zooming into it.

A fractal is a point set that has a dimension which may be non-integer (Mandelbrot 1983). The most famous fractal set was named after its discoverer Mandelbrot (Fig. 2). To define it, we need complex numbers, which I hope the reader is half-way familiar with. Otherwise just skip the definition.

Let *c* be an arbitrary, but fixed complex number; let z_0 be another complex number. We take it to start the iteration $z_{n+1} = z_n^2 + c$. Depending on the chosen value of *c*, the iterated sequence of z_n may, or may not, converge when *n* approaches infinity. Those points *c* in the plane of complex numbers, for which the sequence converges, become members of the special and remarkable set called the Mandelbrot set (or M-set).

The M-set has been explored in many of its properties. Hundreds of images have been produced showing those properties. Figure 2 demonstrates a process of zooming into the M-set. The set itself appears to the upper left. It shows a little rectangle whose contents appear enlarged to the right. There, another rectangle identifies the frame for the next enlargement (middle row, left), and so on.

Now consider what we are witnessing here. Given is a simple formula. We use it to evaluate convergence or divergence of the sequence defined by it for a given value *c*. Depending on the result, the point *c* is included in the M-set, or excluded from it. The shape of that set is intricate. But most tantalizing are its inner properties that the computer can help to reveal.

You may not like dealing with complex numbers, but you will agree that the iterative formula looks simple. All it requires is: take the number calculated last, square it, add the constant *c*, and repeat; do this indefinitely in order to discover convergence or divergence of the sequence. A program for exactly this schema is written within a few lines. However, to do so, one item is missing: Convergence cannot appear in finite time. Therefore, the iteration must be stopped after a reasonably long time. But this pragmatic measure may result in errors. The image is, therefore, only an approximation of the M-set. Aesthetics and algorithmics? An extremely simple algorithm creates a rich aesthetics of ever-changing visual appearance. Is this okay?

4.2 The Rendering Equation

The rendering equation is one of the most powerful devices in computer graphics, as you see in Figure 3. Published in 1986, it is one of the few cases that describe precisely the task a difficult problem in computer graphics must solve (Kajiya 1986).

Let a three-dimensional scene be given as a set of volumes and surfaces together with light sources illuminating the scene. The equation describes the intensity of light reflected at point *x'* in direction of point *x*. It is the sum of two terms. The first is the light emitted at *x'* and sent to *x*. The integral expresses all the light reflected at *x'* in direction to *x* that gets to *x'* from anywhere else in the scene. There is no other way for light to leave *x'* than one of these two. The multiplicative term $g(x, x')$ expresses the visibility of *x* when viewed from *x'*.

$$I(x, x') = g(x, x') \left[e(x, x') + \int_S r(x, x', x'') \, I(x', x'') dx''\right]$$

Figure 3: The rendering equation of computer graphics.

If the rendering equation can be solved for a concrete class of scenes, the final step of projecting the scene and rendering the image is simple. The well-known method of ray tracing, for instance, constitutes a solution under further restrictions.

The aesthetics of this difficult algorithmic problem are not of a directly sensuous character. It is a very complex situation, the mental sensation of collapsing into one expression. This is an intellectual joy, even if practically not so overwhelming.

4.3 The World of Numbers

We all believe to know fairly well what numbers are. But do we think of more than the natural numbers like 1, 5, or 1000, or of the fractions like 1/2 or 7/8? Mathematics knows of numbers as an hierarchy of classes of numbers where one is contained in the next class.

We know the natural numbers best, because we count 1, 2, 3, to infinity. If we know one of them, how ever large it may be, we can say what the next one is. Kronecker thought the natural numbers were made by God, while everything else was human creation (Bell 1986).

The integer numbers are the natural numbers, plus the zero and the mirror images of the natural numbers with negative values. The rational numbers like 3/5, -6/13, or 4/7 contain all the integer numbers plus all pairs we can build from two integers. They are the fractions. They allow subdividing wholes into pieces of equal size.

Already at this point, one of the mental miracles of mathematics appears. Clearly, there are more integer numbers than natural numbers. It seems there are twice as many. To each positive integer, n, corresponds the negative integer $-n$. But both of the positives and the negatives, there are infinitely many. Their infinity is called 'countably many'. Call it a. It has the property that $a + a = a$. This appears weird and insane to the common mind, because it cannot be true in the finite world, unless $a = 0$.

Even stranger may be the fact that the rational numbers between 0 and 1 are still countably many, although they are somehow $a \cdot a$. This product turns out to be only countably many, a, again.

A big jump occurs when we embed the rational numbers into the real numbers. They can be thought of as the decimal fractions with infinitely many decimal places. Already when writing 1/3 as a decimal fraction, we need indefinitely many places: $1/3 = 0.3333....$

But the reals are so many that they can no longer be counted. We say, there are continuously many of them. The proof of this marvellous fact is only about 150 years old, a recent insight.

Usually, the hierarchy of numbers ends with the complex numbers (Fig. 4). They can be taken as pairs of real numbers. Each complex number stands for exactly one point of the plane which, for this reason, is called the complex plane. Complex numbers require that you are willing to twist your mind. You may remember square roots. The square root of 4 is 2 (because $2 \cdot 2 = 4$), that of 9 is 3, and square root of 2 is something starting with 1.41421... .

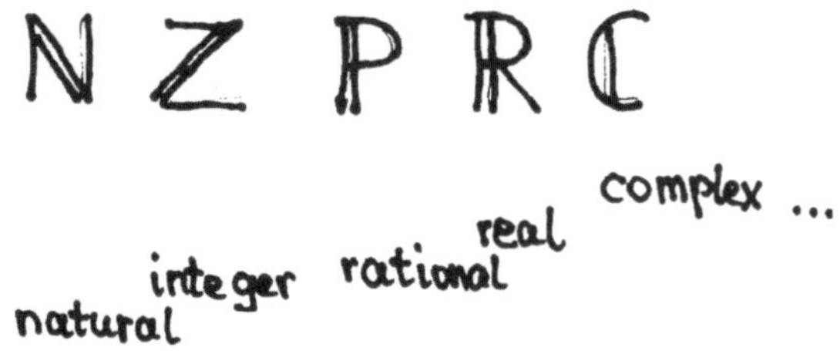

Figure 4: The hierarchy of spaces of numbers.

But what is the square root of -1, a negative number? If it exists, it must multiply by itself to result in -1. We remember from high school that a positive number multiplied by itself results in a positive number, and the same is true for any negative number. So there cannot be a number of the kinds we know which, when multiplied by itself, would produce a negative. To get out of this dilemma, Carl Friedrich Gauß daringly created the one number needed: he introduced the fantastic number i with the miraculous property that, if multiplied by itself, yields -1. This single building block (called the imaginary unit, i) is enough to model all square roots of negative numbers.

This must be enough for a brief tour through the world of numbers: the entities we use for calculations. They are inventions, beautiful creatures of our mind.

4.4 Infinite Series

What is the sum total of adding up 1 + 1/2 + 1/4 + 1/8? After a short moment, you probably say, well, that's the same as 8/8 + 4/8 + 2/8 + 1/8, so it is 15/8, a little less than 2. The trick you do is that you re-name the given numbers so that they have the same denominator (here: 8). With this, the addition is trivial. But what if you now ask for the sum total of something completely different:

$$1 + 1/2 + 1/4 + 1/8 + \ldots$$

where by the mysterious dots you mean that the addition should be carried out indefinitely. The task seems to be impossible. How can you continue adding up infinitely many numbers? We enter real mathematics. For now a task has to be solved that first needs a clear definition before it makes any sense at all. In this particular case, we may apply a trick. That trick is the aesthetics of the situation. We call the still unknown sum, S, even though we don't know yet whether such a number S truly exists. But if it does, we can conclude that, since

$S = 1 + 1/2 + 1/4 + 1/8 \ \ldots$ it must also be true that $S/2 = 1/2 + 1/4 + 1/8 + \ldots$.

Subtracting the two (formal) equations yields

$$S - S/2 = 1$$

because with the exception of the first element, 1, $S/2$ contains exactly all those numbers that S contains, too. The subtraction wipes them out. Now standard algebra tells us that

$S/2 = 1$, and, therefore, $S = 2$.

Is this not totally fabulous and mind-blowing? The result tells us that the infinite series above has a sum in finite terms, and that sum is just 2. Surprise. If this is true for this particular series, something similar may also be true for other cases of infinite series. Second, the example shows that at times a bit of tricky thinking helps and may be enough to find the solution. Isn't the pleasure we draw from this of aesthetic quality?

4.5 An Algorithm for Sorting

Given is a finite sequence of numbers in random order, like {7, 18, 6, 3, 15, 3}. They must be re-arranged in ascending order. Figure 5 sketches a recursive algorithm to solve this task. It is written in a form similar to program notation.

One strategy to solve the task is to determine the maximum of these numbers and make it swap places with the last element of the sequence. In our example, 18 and (the right-most) 3 would swap, resulting in the new list {7, 3, 6, 3, 15, 18}. We repeat this procedure with one small but important change. Since we have already assured that the largest number is now located at the end, we ignore it and consider only the sequence that is shorter by 1. Its maximum is 15, which we swap with itself since it happens to be placed exactly where it is supposed to be. So the sequence has still the same form, and we continue by performing our basic step on the sequence of only the four first elements.

```
function sort (X, n) is
  { if n ≤ 1 then X
    else { determine the max of X
           and exchange it with the last
           sort (X, n-1)
         }
  }
```

Figure 5: A sorting algorithm in recursive pseudo-code.

In Figure 5, variable X stands for the sequence to be sorted, and n is the number of currently relevant elements in X. The elegance of this recursive formulation is fabulous. It tells us a lesson to take home: Never deal with the entire situation! Only solve the basic trivial case which here is the sequence X of just one or no element. In that case, X is already sorted, nothing is left to do. But if there are at least two elements, improve the situation just a bit by pushing the currently largest number to the current end, and solve the same problem ('sort') for a shorter sequence (of length n-1). Is this not great? Is it not inherently beautiful?

4.6 Three Artists

In this section, I have presented five aspects of mathematics and algorithmics. The selection could have been different in all respects. My intention was to make the reader aware of some aesthetic features hidden in forms, patterns, or utterances that are problems mathematicians deal with from day to day. The elegance, strength, or clarity in those examples is awesome when you consider the complexity of the contents expressed in those patterns.

Figure 6: Harold Cohen 2010 (Source: Courtesy of the artist).
Figure 7: Manfred Mohr 1999 (Source: Courtesy of the artist).

This last subsection turns around our viewing direction but without talking much about it. Figs. 1 (see first page of this article), 6, and 7 are examples of algorithmic art chosen from the vast domain of artistic production with computers. The three images are of considerably different aesthetics. The artists themselves have designed the programs. All three, Harold Cohen, Manfred Mohr and Casey Reas are artists who in their different careers turned to the algorithmic world, learned how to program, and became masters of this new field. Manfred Mohr (Keiner et al. 1994; Herzogenrath et al. 2007) and Harold Cohen (McCorduck 1991) belong to the first generation of artists who, right after mathematicians had started the aesthetic use of computing machinery, did the same with deeper understanding.

Casey Reas is a leading representative of a new generation of 'algorists'. He is one of the constructors of a programming language and system: Processing (Reas & Fry 2007; Reas & McWilliams 2010). In him we see appearing what has been in preparation for about forty years: the union of aesthetics and algorithmics, a new kind of intelligence. Cohen and Mohr belong to those who have proven that this is possible.

What they have shown in their activity and work is the deep entwinement of the virtual and the actual that is required and celebrated in each work of art, be it of algorithmic origin or not. In the algorithmic case, part of the virtuality that is in the work of art is transferred to actuality. This process of actualizing virtuality seems to get developed even more severely and noticeably in FabLab modes of generating works.

Acknowledgement

A great thanks goes to my friends Manfred Mohr, Harold Cohen, and Casey Reas for the permission to use their art in print versions. I also wish to thank the editors of this book for inviting me to contribute to a movement of which my work is not a part anymore. Without the students from around the world who shared their time with me, nothing would have been possible. Decades ago, great discussions with my friend Ludwig Arnold helped me grow the mathematical kind of thinking I have relied on ever since.

References

Baumgarten, AG 1986, 1750, *Aesthetica*, Olms, Hildesheim.

Bell, ET 1986, 'The Doubter: Kronecker', Ch. 25 in *Men of Mathematics: The Lives and Achievements of the Great Mathematicians from Zeno to Poincaré*, Simon and Schuster, New York, pp. 466-483.

Bense, M 1965, *Aesthetica*, agis, Baden-Baden.

Bense, M 1971, 'The projects of generative aesthetics', in Reichardt, J (ed.), *Cybernetics, art and ideas*, Studio Vista, London, pp. 57-60.

Boden, MA 2010, *Creativity and art. Three roads to surprise*, University Press, Oxford.

Flusser, V 1983, *Für eine Philosophie der Fotografie*. Europ. Photographie, Berlin.

Hardy, GH 1967, 1940, *A mathematician's apology*. University Press, Cambridge.

Herzogenrath, W, Nierhoff, B & Lähnemann, I (eds.) 2007, *Manfred Mohr. broken symmetry*, Kunsthalle Bremen, Bremen.

Kajiya, JT 1986, 'The rendering equation', Proc. SIGGRAPH 1986, pp. 143-150.

Keiner, M, Kurtz, T & Nadin, M (eds.) 1994, *Manfred Mohr*, Waser, Weiningen-Zürich.

Mandelbrot, B 1983, 1977, *The fractal geometry of nature*, W.H. Freeman, New York.

McCorduck, P 1991, *Aaron's code: Meta-art, artificial intelligence, and the work of Harold Cohen*, W.H. Freeman, New York.

McCormack, J & d'Inverno, M 2012, *Computers and creativity*, Springer, Berlin.

Nake, F 1984, 'Schnittstelle Mensch – Computer', *Kursbuch* 75, pp. 109-118.

Nake, F 2001, 'Das algorithmische Zeichen', in Bauknecht, W et al. (eds.), *Informatik 2001. Tagungsband der GI/OCG Jahrestagung*, Bd. II, pp. 736-742.

Nake, F 2009, The semiotic engine. Notes on the history of algorithmic images in Europe, in *Art Journal* 68, 1 (Spring), pp. 76-89.

Reas, C & Fry, B 2007, *Processing. A programming handbook for visual designers and artists*, MIT Press, Cambridge, MA.

Reas, C & McWilliams, C 2010, *Form + code in design, art, and architecture*, Princeton Architectural Press, New York, NY.

DIGITAL REALITIES, PHYSICAL ACTION AND DEEP LEARNING

FABLABS AS EDUCATIONAL ENVIRONMENTS?

HEIDI SCHELHOWE

1. Introduction

Humans learn everywhere, all the time. In order to survive, we must constantly be able to change and adapt our behavior according to our (new) environment. FabLabs, too, are places where people keep learning. Even though this seems obvious, it has to be considered which arrangements have to be made when learning is supposed to happen best, especially for young people.

At this point, one needs to consider the meaning of complex and sustainable learning. Complex or deep learning means that not only skills for repeatedly acting according to fixed rules are concerned, but that (in the sense of Piaget's understanding of learning) the alteration of mental models as a change of oneself in interaction with the environment takes place. This is the deeper meaning of the German word 'Bildung'. Sustainable learning means that, according to a (new) mental model, different situations can continuously be handled where the abstract model is applied appropriately. In the modern Western, quickly changing, information-based world, this is the kind of learning asked for. We have to think about suitable settings where this can happen best.

Evaluations of digital media environments and also of educational events in FabLabs often just point out that children or adults had fun, liked it, or were very concentrated and persistent. This is an important pre-condition, but not sufficient when it comes to proving complex and sustainable learning.

In my paper, I want to explore from a general and theoretical point of view the very specific and original benefit that FabLabs, with their not that inexpensive infrastructure, can offer in comparison to other educational environments, beyond the obvious fact that skills can be trained and learners can have fun. Some argue that through the master-disciple relationship, learning can happen without abstract and alienated indoctrination, more by role model, do-it-yourself, and trial-and-error.

This is no doubt a great way of learning and a big advantage, and it is common practice throughout the world, yet does this way of learning not substantiate the specific enthusiasm about learning in FabLabs.

Others claim that people with very different expertise, accessible even over the whole globe via the Internet, meet in FabLabs, and that this can be seen as the major benefit of FabLab learning: a broad and encompassing transfer of knowledge. Indeed, this seems to be a specific benefit of the Internet and virtual (learning) communities. But FabLabs, I will argue, can contribute even more to topical education that has to do with the specific relation between material and mental realities.

I will first reflect on the role of physical action in learning. I will then explore how the computer as the technological core of digital media imposed a rather new understanding of the relation between the material and the mental world that was addressed in the intellectual history of modern Western world as more and more separated. What is the character of the new things as outcomes of a digital instead of a hand-made or regular industrial production? This will be explored in the next section and will bring me back to the leading question of specific learning potentials in FabLabs.

2. The Role of Physical Activity in Learning

For a long time, reform pedagogy has stressed the close interconnection between physical activity and mental understanding for learning. The German word 'begreifen' states this connection: 'begreifen' contains the word 'greifen', which means to grasp physically, whereas 'be'greifen says mentally understanding. The word meaningfully expresses that both sides, i.e. physical and mental processes, are inevitably linked. Johann Heinrich Pestalozzi (*1746 †1827), one of the first educational thinkers who accentuates the idea of self-determined learning (to be able to help oneself), develops learning materials that are supposed to support learning with brain, heart and hand, concept building and activity (Pestalozzi 1927–1996). Friedrich Fröbel (*1782 †1852), who coined the term 'Kindergarten', is the educator who brought to mind the relevance of play for learning: "Play is the highest phase of child development, the human development at this stage. It is the free-acting presentation of inner reality" (Lange 1966, pp. 33-34; own translation).

Fröbel-gifts or Fröbel-materials are designed to inspire the child in his/her entirety, to activate emotion, reasoning, motor skills, fantasy, and creativity. Orb, ball, bowl, cube, embroidery belong to these materials, amongst others, and they are also used to experience contrariness – the soft, inaudible ball against the hard, sounding bowl, the round orb against the edgy cube. The materials are supposed to evoke experiments and insights into the world through own activity. Development is seen as the result of own productive activity.

Especially in Maria Montessori's (*1870 †1952) approach, materials get a central status. Materials should be designed in such a way that they are of a stimulating nature and challenge the child to challenge his/her intellectual development.

Maria Montessori formulates the following requirements for material:

- It must attract attention.
- It is characterized by simplicity.
- It allows by itself feedback for the child, so the child can see his/her progress without instruction by adults.
- 'One property' should be addressed in order to attract interest and maintain concentration.
- The 'entirety' of a property that refers to a bigger entity has to be existent in the material; general principles have to be represented in the material (Montessori 2001, pp. 112-113).

Pestalozzi, Fröbel and Montessori, as well as Vygotsky, Freinet or Dewey see the prospect of child development in the fact that he/she constructs knowledge by him/herself through physically manipulating his/her environment. All these educators stress that body and mind build an entity and have to be considered in their role in coaction of acting and reflecting. In Deweys words: "The question of the integration of mind-body in action is the most practical of all questions we can ask of our civilization" (Dewey 1984, p. 29).

Jean Piaget, who, based on his expertise in developmental psychology, is a visionary of constructivist learning, laid scientific ground for the conviction that knowledge is not stored, but constructed through actual activity. In his studies, he proves that important changes in children's world models occur through acting with objects. Information is gathered and interpreted according to own world models. When contradictions occur during practical, physical activity, the own models are challenged and have to be changed, which is when complex learning happens. (Piaget 1974)

Piaget suggests that his findings in respect to the role of physical action only apply to children. He defines learning as a continuous development from concrete to abstract thinking. Whereas the child needs the handling with concrete physical objects in order to experience the limits of his/her mental models, Piaget assumes that older children and adults think in abstractions. Amongst many other researchers, Sherry Turkle and Seymour Papert (1990) have criticized this understanding as an overemphasis of abstract thinking against concrete acting that plays an important role in the development of all humans, also for adults. With the name Lifelong Kindergarten, the team of Seymour Papert and his successor Mitchel Resnick at MIT Media Lab made this principle of learning their mission: "We are inspired by the ways children learn in kindergarten: when they create pictures with finger paint, they learn how colors mix together; when they create castles with wooden blocks, they learn about structures and stability. We want to extend this kindergarten style of learning, so that learners of all ages continue to learn through a process of designing, creating, experimenting, and exploring" (Resnick 2012).

3. The Disappearance of Objects and Physical Action in Learning

In the Industrial Age, the conditions of production have changed towards mass production, big manufacturing plants and machinery that are in the hands of equity owners. As the result of this industrial change, children were protected from working under these industrial conditions. Learning by doing and following role models as it happened in the workshops of handicraft production was no longer possible under the new regime of industrial plants, which not only were extremely dangerous for children, but also came together with de-qualification and alienation of workers from their work and the product, as it is impressively described by Karl Marx (1864). Industrial plants cannot serve as places of education; instead, they are spaces for exploitation of labor.

Competences could no longer be acquired incidentally by acquisition of knowledge in the basic production arenas of the time; hence the transmission of competences was evacuated to specific institutions, separated from production. At the same time, general and more abstract requirements like being able to read, to write, and to calculate developed out of conditions in modern societies. The acquisition of these qualifications was taken out of contexts through specific professional teaching as abstract procedures, conveyance of information in favor of self-acting appropriation. This characterizes schools as institutions in industrial societies even up to today.

With post-industrial developments and the dawn of information or knowledge society, work in more and more fields develops towards working with and on signs, whereas manual work is reduced and done by machines (and in developing countries). Digital cultures are associated with this development, handling on and with signs and alienation from body and the material environment. Some educators and scientists claim that schools and parents should rather have their children make 'original' physical and bodily experiences and therefore should prevent them from extensively using computers at an early age. Spitzer (2012) recently published a bestseller in Germany with the title Digital Dementia (Spitzer 2012; own translation).

This suggestion misses the needs of young generations and also disregards people's necessity to relate to their society and to understand recent developments in order to grow and to be able to act as modern citizens and workers. Computer technology and the Internet imbue the recent world. There is no need for escape, and the virtual reality is just as real as the material world.

4. Computers and Digital Culture

What makes computers so powerful and encompassing to shape and to promote new societal and cultural realities? Alan Turing invented the basis of modern computer technology as a mathematical and purely semiotic concept in the 1930s. The exciting and provoking aspect of Turing's idea was to define calculability, a mental concept, as a means of a physical entity, a machine, as it had been known only in the world of production of material goods until then. Andrew Hodges (1992) in his Turing biography elucidates how the passion to explore connections between the physical and the mental world drove Turing's research. His groundbreaking thought (following Hilbert) that mathematical processes could be seen as just mechanical processes to be simulated and finally replaced by a machine, revolutionized labor as well as private life and the ways of thinking about mental processes. Defining calculability by a machine, Alan Turing 'constructed' with his abstract machine the possibility to apply Taylorism – the tremendously successful concept to rationalize physical work – to cognitive processes. (For further information about mathematical models of computers see Nake in this book). When Konrad Zuse then in the 1940s applied for the patent for the first electronic computer he had constructed, he argued that his computation machine was an "incarnation of mathematics" (Zuse 1993, p. 100; own translation).

The separation between the material and the mental world characterized the intellectual history of the Western world. The invention of the abstract machine by Turing was the dawn of a new perspective on this relation. With 3D printers and their penetration of everyday life, this new relation can be put in a nutshell as follows: machines can capture your ideas and convert them into something material without further interference of human action. Thoughts and ideas can be made into products by clicking a button. Before then, such a process was only believed to be possible through witchcraft, when a spell could change physical appearances and put things into movement.

5. New Things – Digitized Information and Materiality

Modern computers, called digital media, hide their core as calculation machines. Instead, the interface becomes the actual essence, visible and accessible for non-technical users and actors. Until recently, this meant that the digital culture is identified with interaction through signs, symbolic manipulation and a culture of immateriality. The capacity of computer programs to initiate the production of physical goods took place behind the walls of industrial plants and did not become part of everyday practice. With the rise of low-cost 3D printers and cutters, which also led to the FabLab movement, this is going to change.

Industrial products do not tell the user anything about how they are produced, by whom and where, or which visions, concepts and models brought them to life. In handicraft, this is different, as the products have an aura, as Bruce Sterling (2005) puts it, and let us guess their origin and their meaning.

Computers are first and foremost products of industrial production. But their software has the very specific characteristic that it is less a product than a process. Using computers means that the process of production is always present through the code processing during the usage. This is in normal circumstances hidden from the user. He/she interacts as if this process was just 'natural' and not caused by human (software-engineer) endeavor.

Bruce Sterling points out this particularity when he speaks about "SPIMES" (Sterling 2005; his spelling). Spimes are "manufactured objects whose informational support is so overwhelmingly extensive and rich that they are regarded as instantiations of an immaterial system" (Sterling 2005, p. 11). These kinds of objects are the outcome of scientific investigation, of model building, and algorithms manifested in computer programs. Models, as Sterling suggests, are more open and flexible than other physical objects known before. The models are more fluid, less static. They attract and evoke creative thinking all over the world. People want to contribute, change, develop further, mash up their ideas with the existing ones, they want to get to know the people who create and use them and want to discuss the ideas that helped them improve and further develop things (Sterling 2005, p. 106). The history of these objects can be tracked in (digital) archives. Hence, errors and mistakes can be uncovered and revealed as well.

6. FabLabs as Learning Environments

Based on these considerations, I see five good reasons to develop FabLabs as relevant places for learning, with a very specific potential for 'Bildung' in the digital era:

6.1 Combining Physical Activity and Abstract Thinking

The importance to interconnect physical activity and abstract model building in the process of learning has been laid out in the second paragraph. In FabLabs, learners can get both: the abstract model is the immediate cause for the physical output. The learner can express, explore and refine his/her idea and finally has to describe it precisely. The learner furthermore has to build the abstract model in his/her mind, splendidly supported by the visual representation of a 3D program that allows for trial-and-error and direct feedback as approach. The product is printed by a machine that can be observed in its process of putting layers of material one on top of the other, obeying the abstract description of a (self-made) computer program.

For educational activities in FabLabs, the challenge is to build and to maintain a strong connection between a physical product and an abstract model. School children, as Paulo Blikstein put it in his talk at the Fab*Education conference in June 2012 in Bremen, Germany[1], may be interested in printing just one and the same keychain over and over again for their many friends and relatives, as its production code could easily be found in Thingiverse or on a comparable platform. This does not suggest complex and sustainable learning though. (See also Blikstein in this compilation.)

Such an approach empowers children as 'masters' of industrial mass production, but learning between physical activity and model building does hardly happen. Thus, we have to make sure that for learning purposes in regular circumstances, both sides are present: the model building process as well as observation and handling with material objects.

1 | For the detailed program and video recordings see the Fab*Education website, viewed 20 October 2012, <http://fab-education-bremen.blogspot.de/search/label/Symposien>.

6.2 Revealing the Model Behind the Scene

Seymour Papert and Sherry Turkle (1990) already elucidated that computers can be handled very concretely, but at the same time they are a climax of the abstract thinking and models, which are still and always inherent in its processes. The abstract models can be made accessible through specific educational designs. Papert was the very first computer scientist proposing computers for children's education. He coined the term constructionism, based on Piaget's constructivism, which means that children rebuild computer programs through own construction, accessing the models through physical and cognitive activity.

Based on these assumptions, the challenge for designers of learning tools and learning environments is to arrange these surroundings in a way that the abstract model that drives the concrete appearance of an object becomes visible and transparent, and invites to understand and to explore it in interaction.

This will imply deep and complex work on the 3D environments actually used in FabLabs. They will need to be adapted to and implemented for educational purposes in the sense of following a "reflexive experience design" (Schelhowe 2012, p. 253 et sqq.). The arrangement of the whole setting has to reflect how the material products and the observation of the production process could be better used in constantly displaying the references between objects and models.

6.3 Initiating Processes of Reinventing and Refining Own Ideas and Products

When the product is printed, it evokes further physical activity like composing parts to a bigger entity or painting and decorating. It even invites the builder to new kinds of construction activities, to enhance the product with microchips, sensors and actuators in order to bring them to 'life'. It invites to go back to the virtual model, to change and to re-fine it. Educators have to make arrangements to provide material and educational settings where this evocation, this challenge to continuously explore new opportunities and new insights, is likely to happen.

This means that FabLabs have to provide a broad panoply of different materials, additional to those that are needed by the printers and cutters, that is, decoration materials of all kinds, tool kits, instruments, microcontrollers such as Arduinos, sensors, motors etc.. An exhibition of a carefully chosen range of types of objects produced in a FabLab should accompany the constantly visible exposition of photographic representations of existing products as well as of their code. This can support imagination and continuous refining and reinventing.

A great potential in learning is to learn not just from success, but also from mistakes. Encouraging learners to look at mistakes made by others and by themselves and thus to encourage them to overcome these mistakes is an important source for FabLab education. We should therefore have a look on weaker examples instead of only considering the best resulting products, in order to support and encourage refining processes.

Sometimes the obstacle to start a production from scratch is a high barrier. It might be easier for beginners to just start with an existing code and to refine it. To download a code from Thingiverse might be just the "low floor" (Resnick 2005, p. 118) that gives entrance to the "high ceiling" (Resnick 2005, p. 118) of highly creative thinking and production that Resnick and others propose for educational toolkits.

6.4 Relating to Post-Modern Society's Conditions

The materials suggested by Maria Montessori and others mostly address children in primary school age. In referring to 'general principles', they relate to rather basic competences like reading and calculating and their occurrence in modern or pre-industrial societies. However these basic materials cannot address present-day complex processes in society and in production. Célestin Freinet (1995) with his printing machine for writing texts in schools has made one of the most approximate approaches to developing a material learning device of modern technology and industrial production.

Being accessible for everyone, FabLabs can open a real avenue for insights into how post-modern production works. They are environments for 'Bildung' in the sense that individuals cannot only develop their personality but also their relation to the world and to society. In combining a very practical approach with complex theoretical insights, FabLabs incarnate accesses to understand the today's world. This is what Ann and Mike Eisenberg called "tangible expressions of important ideas" (Eisenberg & Eisenberg 1999, p. 260), when they spoke about using robots in education, or when Murray (2003) said about digital media, "the digital medium is as much a pattern of thinking and perceiving as it is a pattern of making things. We are drawn to this medium because we need it to understand the world and our place in it" (Murray 2003, p. 11).

This thinking is really different from the traditional tinkering and handicraft activities that modern pedagogy suggests and some schools have adopted. In FabLabs, we should not waste this specific potential, but fully stress and use it. The big processes that happen in modern production in plants and in everyday surroundings can be simulated and carried out in smaller and simpler grades, but the principles are mostly the same. In order to gain these insights and to relate oneself to society, we have not just to counteract the maker community to industrial production, but we also need to refer to it, to mention these processes and to increase the awareness of how knowledge and competences are gained in FabLabs.

FabLabs can also serve as spaces where competences that are needed in the labor market – especially in the STEM (science, technology, engineering and mathematics) field, but also so-called key competences – can be adopted. FabLabs as educational labs could also be places where processes and products from 'big' productions are exhibited and associated with FabLab productions.

FabLabs have the potential to become high-level educational centers where practical experience leads to complex understanding without the physical, mental, and social set of problems and dangerousness of industrial plants.

6.5 Social and Community Learning

In his groundbreaking socio-constructivist approach, Lev Vygotsky (1926) emphasizes the importance of cultural mediation and interpersonal communication for the development of the child. He stresses context, community, and social interaction as basic conditions of physical and mental growth.

In FabLabs, learners meet communities of creative people with very different backgrounds and knowledge who help and inspire each other, and where learners can integrate. FabLabs also connect to network communities, where Jenkins locates the dawn of a new "participatory culture" (Jenkins 2006) that characterizes, in his point of view, the young generation. "Play, Performance, Simulation, Appropriation, Multitasking, Distributed Cognition, Collective Intelligence, Judgment, Transmedia Navigation, Networking, Negotiation" (Jenkins 2006, p. 4) are characteristics that FabLabs connect to the interests and needs of youth in a digital culture. Hence, FabLabs can consider the individual child's developmental status and guide every one according to his/her own prior knowledge and education, thus not asking too much of one. FabLabs have to understand how to meet these needs of youth of all genders, ethnicities, and educational backgrounds and have to overcome the social and digital divide.

Papert always emphasizes the importance of personally meaningful objects to be produced in his workshops. This is also true for FabLabs. Following pre-fabricated plans will not motivate learners. Instead, they have to regard their products as more than just an ordinary object by making them personally meaningful. Learning more about social regulations to be applied in order to unfold the potentials for inclusion and empowerment – for the physical room as well as for the web community – will be a crucial task when FabLabs develop towards educational labs.

7. Conclusion

FabLabs are exciting new learning environments, and fabbers experience learning in all the depicted dimensions all the time without thinking about being 'educated'. If we want to ensure that complex and sustainable learning can happen for more people, bridging the social and digital divide, beyond just addressing those who possess self-motivation, we have to think about possible changes in conditions and arrangements.

Good designers know how to design educational objects that by themselves evocate implicitly the challenge of learning without emphasizing explicit instruction. And good educators know how to arrange situations where instruction can be minimized. Good educational objects or environments evoke self-generated questions, give feedback on errors and allow for learning from mistakes. They focus on general insights and abstraction that allow for transfer of the newly acquired knowledge to many other situations. All this is present in FabLabs much more than in many other settings. Its objects are graspable, but they appear as instantiations of models, thus they refer by themselves to an abstraction that represents a more general idea and can be communicated. A stimulus is present to further develop already existing ideas without any fixed end. The history of objects as well as the mistakes are stored and can be accessed in the code, and are open for generic learning about objects. The FabLab environment opens plenty of opportunities for social and participatory learning, whereas schools as institutions of industrial society stay behind or make it at least difficult to offer innovative potential for new forms of learning.

In my paper, I addressed some of the potentials that FabLabs as very special spaces offer for education. I tried to suggest some measures that have to be taken in order to make the best of FabLabs as environments for complex and sustainable learning, not only for 'nerds', but for everyone. As an enthusiastic supporter and member of the FabLab community, I would like to contribute to a deeper understanding and practical proving of what learners need and how we can arrange tools, surroundings, and the wider context in a way that learning can happen best.

References

Dewey, J 1984, 'Body and Mind', in Boydston, JA (ed.), *The Later Works: 1925-1953. Vol.3: 1927-28*, Southern Illinois University Press, Carbondale, pp. 25-40.

Eisenberg, M, Eisenberg, AN 1999, 'Middle Tech: Blurring the Devision between High and Low Tech in Education' in Druin, A (ed.), *The Design of Children's Technology*, Morgan Kaufmann, San Francisco, pp. 244-273.

Freinet, C 1995, *Die Druckerei in der Schule*, Schriftenreihe des Förderkreises Schuldruckzentrum Pädagogische Hochschule Ludwigsburg, Ludwigsburg, Germany.

Hodges, A 1992, *Alan Turing: The Enigma*, Vintage, London, UK.

Jenkins, H 2006, *Confronting the Challenges of Participatory Culture: Media Education for the 21st Century*, viewed 11 October 2012, <http://digitallearning.macfound.org/site/c.enJLKQNlFiG/b.2029291/k.97E5/Occasional_Papers.htm>.

Lange, R (ed.) 1966, *Friedrich Fröbels gesammelte pädagogische Schriften. Erste Abteilung: Friedrich Fröbel in seiner Entwicklung als Mensch und Pädagoge. Bd 1: Aus Fröbels Leben und erstem Streben. Autobiographie und kleinere Schriften. Berlin 1862*, Faksimiledruck, Osnabrück, Germany.

Marx, K 1864, *Das Kapital*, Marx/Engels: Werke. Band I 1983, Dietz Verlag, Berlin, Germany, viewed 17 December 2012 <http://www.mlwerke.de/me/me23/me23_000.htm>.

Montessori, M 2001, *Die Entdeckung des Kindes*, Herder, Freiburg i. Brsg, Germany.

Murray, JH 2003, 'Inventing the Medium' in Wardrip, F & Montfort, N (eds.), *The New Media Reader*, MIT Press, Cambridge, MA, pp. 3-12.

Pestalozzi, JH 1927–1996, *Sämtliche Werke. Kritische Ausgabe. Begründet von Artur Buchenau, Eduard Spranger, Hans Stettbacher*, Gruyter, Berlin, Zürich / Germany, Swiss.

Piaget, J 1974, *Theorien und Methoden der modernen Erziehung*, Fischer, Frankfurt a.M., Germany.

Resnick, M, Silverman, B 2005, 'Some Reflections on Designing Construction Kits for Kids', *Proceedings of the 2005 conference on Interaction design and children*, ACM, New York, pp. 117-121.

Resnick, M 2012, *Mission Statement: Lifelong Kindergarten*, viewed 17 December 2012, <http://llk.media.mit.edu/mission.php>.

Schelhowe, H 2012, 'Interaktionsdesign für reflexive Erfahrung: Digitale Medien für Bildung', in Robben, B & Schelhowe, H (eds.), *Be-greifbare Interaktionen. Der Allgegenärtige Computer: Touchscreens, Wearables, Tangibles und Ubiquitous Computing*, transcript, Bielefeld, Germany, pp. 253-272.

Spitzer, M 2012, *Digitale Demenz: Wie wir uns und unsere Kinder um den Verstand bringen*, Droemer, München, Germany.

Sterling, B 2005, *Shaping Things*, MIT Press, Cambridge, MA.

Turkle, S & Papert, S 1990, 'Epistomological Pluralism: Styles and Voices within the Computer Culture', *Signs: Journal of Women in Culture and Society*, no. 1, pp. 128-157.

Vygotsky, L 1997, *Educational Psychology*, St. Lucie Press, Boca Raton, Florida.

Zuse, K 1993, *Der Computer – mein Lebenswerk*, Springer, Berlin, Germany.

A UNIVERSE OF OBJECTS

CORINNE BÜCHING

Figure 1: Spectacles and decorating material
(Source: Photography by Justus Holzberger and Corinne Büching).

1. Introduction

Technical revolutions have been occurring and changing society for several decades: The invention of Personal Computers (PCs), the World Wide Web, the Web 2.0, followed by the development of smartphones and cloud computing. We seem to be at the peak of digital penetration of our daily lives. Aided by sensors, actuators and micro controllers everyday digital technology – GPS systems warning us of traffic jams or radiators regulating temperatures according to day time – will intercommunicate even more in the future. The German newspaper Frankfurter Allgemeine Zeitung (FAZ) claims that "the programmable world is coming up" (Schmidt 2011; own translation) as machinery is increasingly going to be interconnected, filling the gap in those fields which yet remain untouched by digitization. Hence, the Internet/Universe of Things was born and a new revolution is on the rise.

At the same time rapid prototyping and personal fabrication are being expanded, leaving their production settings behind and extending into our everyday lives. Due to these developments virtual data from complex technologies seem to be materializing magically. This is nothing new in industrial production, where IT, digital design and material production were joined a while ago. This interconnection can now also be found in FabLabs, paving the way for its spreading into everyday life. Things are essential in FabLabs and symbolize the new connection between 'the world of the intellectuals' and the material world. This connection stands for the beginning of a new era, in which the machine that decodes the signs also controls material production while creating new objects. It is a place where objects do not remain abstract ideas or mental concepts but become actual material products, without human beings even needing to use their hands for it. This type of 'one-step' production is already common in industrial production and is also going to become more meaningful in everyday life (Boeing 2010; Neef, Burmeister & Krempl 2005; Gershenfeld 2005).

Therefore, the website Thingiverse, which is an essential part of the FabLab movement, foresees a Universe of Things. On this site one finds things and tools of all kinds to be printed out at home: digital design for real, physical objects, such as comic figures, jewelry, games, and furniture. There are no limits to your imagination on Thingiverse and everyone is invited to use, alter, develop or personalize every creation, using specific software. Thanks to fabrication devices, virtual things can easily be brought into the material world, directly into the living rooms. Personalized things, self-made products and individual creations are gaining importance as they support and symbolize processes of individualization. Besides Thingiverse, there are a number of other websites – as part of the maker culture that help pave the way towards a rethinking of personalized objects – where self-made and individualized products can be displayed and sold (for examples see the article of Katterfeldt in this compilation). Not only do the fields of production change, but the fabricators/inventors are also becoming more independent in terms of the production of material objects. This change suggests a new form of connection with the industrial (i.e. professional) production sector. The production of personalized, physical objects through interaction with the digital media of digital fabricators in private households is not very common yet, however, it is likely to spread once 3D printers have made their way into our homes. FabLabs are places where these changes in the personal production world and its common social practices can be observed. They are places where every one of us can turn something digital into something material, giving the computer revolution a new quality, as virtual things are entering the actual, physical world.

Even within the sciences these changes in the means and meanings of production and objects have been realized. Social sciences and humanities are increasingly dealing with the influence of everyday objects on human life, questioning the dualism between subject and object, which was axiomatic for a long time. Furthermore, post-structuralist approaches especially established and promoted by the work of Bruno Latour (2005) and others that include a material perspective on the world become more and more popular.

Things that are able to intercommunicate as well as virtual things that become physical are popular in the present discussion about things. This is reason enough to have a closer look at the things, which are being developed in FabLabs with the aid of new production techniques. Where exactly does the fascination of DIY things – apart from the traditional knitting, sewing and repairing – come from and what is so tempting about 'digital crafting'? How does this change our view of these objects and what is new about the relation between materiality and virtuality?

In order to answer these questions, I will introduce empirical data and conclusions following a theoretical introduction of the Universe of Things. Key words that will be dealt with in this chapter include: the Internet of Things, ubiquitous computing, the dualism of subject and object and agency. Subsequently, I am going to explain the research setting of a workshop named 'Shape your world in FabLabs'. Empirical methods of object research, picture analysis as well as interview interpretation will be used to introduce and analyze the things that were made in the workshop, while at the same time always relating back to relevant theoretical links like reverse engineering and the relation between virtual and material world.

2. Digital Things – Digitized Objects

The lines between the virtual and the physical world are beginning to diffuse. Computers are hidden in numerous everyday objects, broadcasting data and even communicating with each other. In as early as the 1990s Mark Weiser (1991) already envisioned such scenarios. He coined the term ubiquitous computing, which means the penetration of everyday life by nearly invisible, intelligent things. As Neil Gershenfeld (1999, p. 57) points out, many technical projects aim at improving technology so far that it becomes invisible. This is increasingly happening and reaching a new quality. The connection of things and their intercommunication make them become more and more important and autonomous in our society. Library books that are equipped with RFID chips, geo-caching objects that are equipped with GPS sensors or postal packages that can be tracked online indicate the beginning of the development of the Internet of Things. Inflexible, installed systems such as power outlets, cords and cables are increasingly being replaced by more modern and flexible inventions. Technologization and digitization are ongoing, thus also changing the perception of these objects which surround us in daily life. Things like localization technologies will make it easy to find lost objects or the objects may even be able to tell us their location themselves. Smart things may make life easier, using IT components – possibilities are endless.

As Mattern (2004) puts it, "the consequences of such a deeply rooted integration of information technology into our everyday life as propagated by Ubiquitous Computing cannot be foreseen. However, if ordinary things know which other things or persons are nearby and what has happened to them in the past, and if they are able to communicate this knowledge with other objects, the consequences will certainly be of great economic and social importance" (Mattern 2004, p. 4; own translation). Controversial discussions focus on the danger of technologies' unreliability and of human dependence on technologies as well as on data security questions.

Our environment is increasingly being penetrated by digitized objects with which human beings interact independently: numerous intelligent automated machines are used on a daily basis ('Embedded Computing'), intelligent clothing enables us to get safely through the night on our bike thanks to LEDs ('Wearable Computing') and objects that are integrated into the environment ('Sensor Networks') make them graspable with all senses (Mattern 2004). Physical, material things with a 'digital origin' that can interact individually with their environment due to their programming surround us every day. What is special about these kinds of objects is that they evoke a new form of contact, the so-called "Embodied Interaction" (Dourish 2004). On the one hand, Embodied Interaction is based on everyday experiences with common material objects and social relations, making use of that knowledge for interacting with 'intelligent objects'. On the other hand, we will also make new experiences with these digitized objects, which will reveal the programmed artifacts' added value (Schelhowe 2007, pp. 51-53). As Mattern (2004) explains, "smart objects and sensor-equipped environments store large amounts of partly private data in order to be able to provide their services at any time" (Mattern 2004, p. 4; own translation). The flow of data, which is inherent to the intelligent objects, is highly sensitive. Smart things are even able to intercommunicate without any human interference. Therefore, Max Weber's long time axiomatic proclamation about human beings being the only ones who can act rationally and reasonably is now questionable. Hence, one can say that this is how smart things achieve some sort of (active) power or agency (Rammert 2007). Can things act? This is one of the crucial questions based on Werner Rammert's and Ingo Schulz-Schaeffer's (2002) same-titled book. The objects' agency varies, depending on the numerous constellations of interactivity between human beings and objects (Büching et al. 2012). In other words: In a "lifeworld" (Schütz 1974) that is increasingly being penetrated by digitization, different constellations of "actors" (Latour 2005) can negotiate in mutual chains of action. Humanities' view of things has changed, not least since the introduction of Bruno Latour's ideas. The dualism between object and subject is increasingly being resolved. The technologization of everyday life also includes an increased number of non-human actors (Latour 2005). Hence, the object's status is being focused and strengthened. Smart things are even able to intercommunicate without any human interference.

In a culture, which is shaped by digitalism, the lines between virtual and physical worlds are blurred. We have been able to "shape our digital environment but not our [actual] physical [one]" (Gershenfeld 1999, p. 78) for over a century. While it was only possible to either write on a computer or by hand, either to go outside to play or to play on the computer, these 'problems' are about to be solved, as Neil Gershenfeld already foresaw in the late 1990s. He postulated that the barrier "between digital information and our physical world" (Gershenfeld 1999, p. 15) must be torn down. Instead of replacing reality with a new artificial one, Gershenfeld (1999, p. 15) suggests that we should develop technologies to improve the existent one. Describing his own first experience with a self-made physical object, Gershenfeld explains that it was one of the most frightening moments for him when he "opened the door of a 3D printer and took out a part that [he] had just seen on the screen of a computer" (Gershenfeld 1999, p. 78) to discover that the object that he had just only been able to see on a screen had materialized into a real object. This process crossed the line between the computer and the outer sphere. Gershenfeld further explains that to him the 3D printer is the mediator between the virtual and the physical world.

Sterling describes such objects in slightly more abstract terms, calling the things constructed in FabLabs "SPIMES [that] are manufactured objects whose informational support is so overwhelmingly extensive and rich that they are regarded as material instantiations of an immaterial systems" (Sterling 2005, p. 11; his emphasis). He further states that fabrication devices can be compared to the "Philosophers' Stone", spimes being the objects that are used to 'create' the future (Sterling 2005, pp. 102 et sqq.). However, Sterling's description should be understood as a form of analytical approach towards such objects rather than as a means of analysis. It is only a matter of time that fabrication devices and spimes will become common. They are increasingly making their way into our world, changing the constitution of society. Machines for production can be found in FabLabs and can even be affordable for private households. Its increasing distribution is symbolic of the new connection between the physical and the virtual world, making this cultural phenomenon visible.

Nonetheless, this interconnection between the virtual and the material world, which becomes visible through the everyday use of fabrication devices in FabLabs, has not been systematically researched. The digitized things should also be scientifically reviewed and understood. What makes smart things special? To ask more precisely: Why do the things that are fabricated in FabLabs fascinate us so much? And what does this tell us about human beings? To what degree do people believe such 'materialized ideas' are meaningful? How does this change our view of the objects and what is new about the relation between materiality and virtuality in our world? FabLabs provide a setting where the objects' 'biographies' and their formation from virtual to physical object can be analyzed in order to draw generalizations about the relation of virtuality and materiality.

3. Shape your World – Create FabThings

In order to explore the fascination with and appeal of things, together with Julia Walter-Herrmann I conducted a workshop titled Shape your World in FabLabs. Eleven participants (five girls and six boys) between 19 and 28 years of age took part in the two-days' workshop, which was held with great support of FabLab St. Pauli, Hamburg, Germany. Half of the participants did not have prior knowledge or experiences with digital devices and programming, and none had been to a FabLab before.

The workshop was conducted within the context of the research project Subject Formations and Digital Culture (SKUDI)[1]. In this project, we focused on the subjects and their stance on the technologically permeated world. The term subject formation, in this case, includes all forms of human existence, including one's biography, attitudes, values, etc.. By subject we mean an active, capable individual who creates his/her environment through interaction with his/her fellow human beings (Arendt 1960), eventually creating personally meaningful objects. In the SKUDI project we offered four different workshops, where young adults were supposed to get meaningful exposure to technology (Büching et al. 2013). The research was focused on the subject in all workshops. In Shape your World, for instance, the focus was on the "Homo Fabber" (Neef, Burmeister & Krempl 2005). While Max Frisch (2008) saw the Home Faber as a typical rational ingenious, the Homo Fabber creates his/her own world (of goods) at the computer and prints it out with the Personal Fabricator. This does not mean that the Homo Fabber is as little creative as Walter Faber in Frisch's novel; instead, he/she is rather very creative in the process of designing, since the 3D machine is able to construct very complex objects (Neef, Burmeister & Krempl 2005, p. 105). Hence, the Homo Fabber is much more independent. In contrast to the focus of the research project SKUDI, this paper does not focus on the subjects but on the objects, i.e. the FabThings that the Homo Fabber produced during the workshop.

The theory of constructionism provides a pedagogical approach (Papert 1980, 1987). I developed a workshop design, which considers both the active subjects and the objects valuable for the learning process. Constructionism builds on the principles of constructivism (Piaget 1974). Seymour Papert, who was a student of Jean Piaget's, further developed this approach and emphasized the great meaning of active construction and activity in the learning process. Constructionism focuses on the objects to learn with, calling them "objects to think with" (Papert 1980, p. 11). Within this paradigm, learning takes place through the construction of personally meaningful artifacts. This active construction process makes the constructors reflect on abstract concepts: "From constructivist theories of psychology we take a view of learning as a reconstruction rather than as a transmission of knowledge.

1 | Four German and Austrian universities are part of this interdisciplinary project, dealing with the question how society is changed by digitization in the fields of work (Hamburg University of Technology), communication (University of Klagenfurt) and learning (University of Bremen). These empirical approaches are contrasted with a historical review of the concept of the subject (University of Münster).

Then we extend the idea of manipulative materials to the idea that learning is most effective when part of an activity the learner experiences as constructing a meaningful product." (Papert 1987)

I implemented the main ideas of constructionism in the Shape your World workshop as follows: at the very beginning, participants are invited to go on a phantasy journey in order to find ideas as to what they would like to make/fabricate/produce during the workshop. This first step ensures the meaningfulness of the built objects for the makers. Afterwards, in the construction phase, groups interact and construct things, dealing with the physical, software-based environment. This is the workshop's core phase, in which participants create, arrange, and program things, making a virtual object on the computer. The construction phase ends when participants print out their objects with the help of the laser cutter or the 3D printer, hence having constructed a physical object. The real, physical object encourages reflection, as the "object to think with" (Papert 1980, p. 11) becomes real. After this reflection phase another construction phase may begin, e.g. to correct mistakes, change the design, etc.. Likewise, the construction phase ends with the real, physical object. A presentation of one's work marks the end of the workshop. This phase is important for participants in order to review the whole workshop process and to present their objects to the public.

4. Object Research and Picture Analysis

Pictures can capture two dimensions, but the objects made in FabLabs are multi-dimensional, possess material qualities, and can be experienced through different senses. Approaches of object research are meant to promote the perception, appreciation and the understanding of the object. To put it differently, the functions and meaning of the objects are to be revealed. On the one hand, this is achieved through an object analysis considering categories such as the material and its quality, color, behavior, utility and benefit, design, etc.. On the other hand, the personal meaning and function of the object for a specific person is being retrieved through focused, narrative interviews that reconstruct the objects' biography from the subjective perspective of its designer.

In pre-industrial times human beings were mostly aware of their commodities' origin, their use and production, and even of their disposal. Food was homegrown, clothing was sewn by hand, and craftspeople made personalized furniture (see also Liessmann 2010, pp. 11 et sqq.). Few things dominated life back then, and their biography could mostly be defined from the beginning on. This is not the case anymore, the production process for most consumer goods can hardly be traced back, and the product is also being disregarded once it has been placed in the garbage. The continuous contact with the product in all its different stages of construction, usage, and decay does not exist anymore, and its production is done in factories. Historically, this vanishing of the things' biographies can be said to have started with the rise of the industrialization and is constantly being strengthened by a growing automation and digitization (Liessmann 2010, pp. 13-14). Where the things come from, which machines and persons made them and which materials they are made of usually remains invisible – Every'thing' is simply there (Liessmann 2010, pp. 14-15). To trace the things back in order to make their production visible is the aim of such object research which becomes possible in FabLabs.

FabLabs are considered social spaces in which a result is 'produced' through social interaction and the negotiation of meaning, hence, a qualitative, non-standardized approach is appropriate. The open as well as interpretive character of qualitative research is reasonable, since FabLabs are a field that has not been empirically researched yet. Moreover, besides human actors, digital media is also going to be considered as means of data collection, as they call for action due to their "evocative potential" (Turkle 1984, 2011).

This empirical study focuses on the objects, which were made in FabLabs. Therefore, not only did I use the data from the SKUDI project (gathered guided interviews, participatory observations and video analysis) but also a great amount of pictures for the analysis, according to the assumption that "social worlds are continuously being designed with and through pictures" (Breckner 2008, p. 1; own translation). The analyzed pictures include photographs taken by the researchers and participants as well as screenshots of the programs and the virtual objects taken by the participants. In this paper, however, I will refrain from such a detailed description of the picture and its segments, in order to introduce the objects as a whole. The analysis will reconstruct the objects' connections between their development, storage, and usefulness (Breckner 2008, p. 4) as well as the reconstruction of the media model (Breckner 2008, p. 4). All of these analyses will be part of the overall interpretation of each object, which will be presented in the following chapter about FabThings.

5. Empirical Results about the Universe of Objects

In our FabLab workshop numerous things of personal value were created, as opposed to consumer goods: a QR code seal for one's personal website, a mug with a personal logo, an engraved tray, a sumo wrestler, various eyewear models, a book binding, name tags, a super hero and lots of jewelry and decorating material. The following chapter analyzes some of these things. But before taking a look at the things in detail, I want to introduce two categories: 'objects to download' and 'homemade objects'.

'Objects to download' are such things that are being collected on websites such as Thingiverse, available to everyone. Such self-constructed models are digitized and the data is uploaded for public access in order to exchange ideas and further develop the constructions in cooperation with other users under certain licensing agreements as Open Source. The objects can be materialized using laser cutters, milling machines or 3D printers. However, for the participants of Shape your World, the materialization of objects to download means that they can show little creativity and personal contribution when it comes to programming and constructing. Therefore, the already complete 'objects to download' often become personalized through individual engravings, hence becoming 'self-made'. They exemplify the decentralization of the producing process and the return of production to the place of its consumption (Neef, Burmeister & Krempl 2005, p. 9), and thus cannot be compared with 'common consumer goods'. It is important to mention that they are manifestations of the movement (see also Julia Walter-Herrmann's contribution in this compilation): FabLabs ask everyone to share everything with everyone on Thingiverse.

This chapter focuses on the homemade objects, i.e. the participants' self-constructed, programmed, and developed objects, since the practice of designing things is "to give form, or expression, to inner feelings and ideas, thus projecting them outwards, making them tangible" (Ackermann 2007, p. 2). Things that were self-invented, -imagined, -constructed, and -programmed embody their developers' ideas. They are the product of the mental idea, leaving traces like program codes, and possess the potential to make their creation process re-constructible for all.

The sequential analysis of the object's biography provides information about the genesis of the object and can equally tell us something about its maker: about his/her ideas, feelings, and relation to the things. FabLabs allow making them tangible. The materialization of the things as a sort of expression of a former thought, of an idea subsequently developed, and of the written algorithm is a special moment. The term 'expression' has a two-fold meaning in this case: (1) from an IT perspective it refers to the electronic-mechanical product of a printing process, and (2) from a linguistic viewpoint it is a sensible combination of signs.

The object's biography is completed once it becomes materialized. The file remains and can be edited and adapted, changed and written anew. That is how various physical objects, whose production was based on the same prototype but which were then further developed in different ways, are constructed. For instance, some might become part of bigger machines (see the 'Fab-tast-O-Matic' in Dittert & Krannich's article in this compilation), and others can be digitally enriched through technologies such as Arduino. FabLabs are places where such materials are easily accessible and likely to be used. To sum it up, homemade objects are self-made through and through: from the first idea to its materialization to its construction and production. Once materialized, the final object needs to be regarded in its context of human attribution of meaning (i.e. that of its developer).

5.1 Spectacles and Reverse Engineering

The eyewear project serves to reconstruct the process of the genesis of the physical object to the virtual object, and back to the physical object, using reverse engineering. The developers go to the flea market to get inspired and develop an idea. This market is set near the building in which the FabLab is located. They return with a fashionable model of sunglasses that was industrially made. The glasses are a suitable object to re-model for all group members. As a first step, the participants take a photo of the glasses and digitize it. Afterwards, they construct a computer model of the glasses, using the software blender. They remove the photo's background alongside the cutting edges of the glasses. The lenses of the spectacles and the trapezoid hole between the glasses are being cut out as well in order to construct an identical model. Various computerized tools enable the digitization of the physical object and software provides editing tools. The result is a virtual copy. The final computerized, cutout model of the glasses gets arranged on a given part on the wooden panel to be printed on. In one printing session as many models as possible can be printed. In order to fill the panel ideally, the developers additionally arrange different symbols and forms that can be used for jewelry and decorating material on the panel. These are byproducts of the materialization of the glasses. Five models of glasses are found on the wooden panel. Six lenses are marked with little star symbols, and two are marked with skull-like figures on one lens and a name engraving on the other. Once the panel has been filled, it is treated/processed with a laser cutter. The glasses and the decorating material hence do "not remain in the sphere of ones and zeros, but end in the familiar field of things that are perceivable with all senses" (Neef, Burmeister & Krempl 2005, p. 15, own translation). In the printing process, the digital and the physical world melt together, and the digital things become real material: "Software creates hardware" (Neef, Burmeister & Krempl 2005, p. 15) and the objects once again become objects in the physical world. The constructed eyewear model is nearly identical with the one from the market that was industrially made.

5.2 Super Hero – The Fusion of the Virtual and Physical World

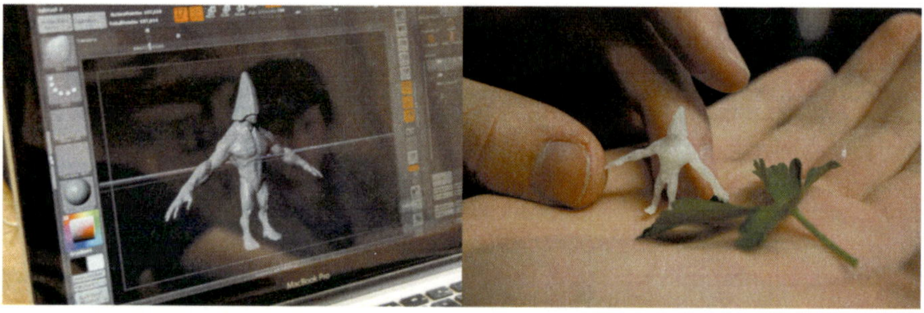

Figure 2: Virtual and physical super hero (Source: Photography by Axel Sylvester).

This picture above illustrates how small the figure described in the following part is. On the palm of a hand, the printed figure was placed next to a leaf. The figure's size is only 2cm and is the portrayal of a super hero avatar. The small size is due to the duration of the printing process. It takes an hour to print a figure of this size with the 3D printer available in the FabLab, hence, it would not have been possible to print a bigger model of the figure due to the workshop's time limit. The Stratasys Dimension 3D printer is a machine which is also used for industrial production for the development of prototypes and the production of small-scale series. The size of the end product is meant to be 15 to 20cm, and the figure is going to be a super hero, which the designer – in the following called "Hero" as a reference to his object – makes for a course in Graphic Design.

In Hero's life "3D stuff", as he likes to call it, plays an important role: "It's just what makes me happy" (Hero, 27 years; own translation). He wants to become a character artist and make a living through the development of such products, for instance, by offering them for sale on websites. Numerous websites with different thematic foci exist, i.e. herobuilders.com, which offers "personalized Action figures that look, talk, & dress just like you" (Herobuilders 2012). Such websites allow customers to personalize super hero figures' heads as well as wooden toy trains and similar toys. Furthermore, they allow the construction of a 3D sculpture out of a 2D drawing, paving the way towards a new industry. In order to achieve his dream job, Hero participates in online courses by renowned and popular artists in his field of interest. He uses the Internet to get access to fields which are not provided at his university and that are beyond his computer science studies. For instance, he deals with disciplines which supposedly are outside his subject area, such as anatomy, in order to be able to better mold the human body and its shape. The courses are "live and allow participants to ask questions and to communicate with the referents directly using an audio chat program, and it really is cool what possibilities the Internet opens up" (Hero, 27 years; own translation).

Hero is fascinated by the overcoming of physical laws in a digital 3D surrounding. Pens and paper are replaced by graphic tablets and modeling tools that offer a huge variety of colors and possibilities for editing. The time-consuming modeling with clay or plastic modeling clay is replaced by 3D modeling and printing. For Hero, another important aspect of the digital work surrounding is the feedback function of digital media. On the one hand, the object itself can quickly give feedback, e.g. if it has the right proportions, if it can stand on an even surface etc.. On the other hand, the object can be presented online either during the process of its creation or after it has been completed. In a very short time, one can thus address an audience of experts who can give the designer important feedback, critique and appreciation: "I learn a lot from them, they give me advice" (Hero, 27 years; own translation). Hero himself is active on those websites, gets inspired by them and comments on other developers' posts. For him, technical news and refinements open up a whole new approach to the world and expand his scope of action. To him, virtual life means a simplification of physical life: "Technologies basically make life easier" (Hero, 27 years; own translation).

5.3 Personal FabThings

Figure 3: Personal FabThings: QR code, personal logo, and engraved wooden element (Source: Logo design by Dennis Herrmann, Photography by Corinne Büching).

Moreover, things were personalized through engravings of names, a personal logo or a QR code. Here we can see a seal with a QR code that refers to a personal website, the logo of a club which was engraved via laser into a glass mug and a self-made wooden element for the arrangement of glasses which has an engraving as well as a personal inscription on the bottom. None of these things can be bought in a store like that, and the "people compose with what is offered by the materials"

(Ackermann 2007, pp. 2-3). They all are unique, self-constructed objects. All objects were constructed out of the wish to produce something that cannot be bought. They are personalized, new objects beyond mass consumption. What is special about them is their personal note which gives these digital things an aura (Sterling 2005). As they were made according to the maker's wishes, they are more appreciated and less likely to be thrown away, because humans tend to cling to self-made things (Anderson 2013).

This fascination with things and their construction begins in childhood, when we build sand castles, caves, games etc.. As children, we were the "makers" (Anderson 2013) of our world. Nowadays, the Internet provides us with a platform to present and share self-made things, to upload photographs and videos, hence making self-made things more valuable in today's society. Anderson (2013) describes the sharing and uploading of things as a normal fact, which is nothing special anymore, because we are used to it nowadays. In FabLabs, this fascination with making can be expanded. With the help of fabrication devices even virtually self-made things can be transferred into the physical world and come to life: "And what is wicked with FabLabs is that if you like something, but you cannot afford it or something, you can make a photo of it, or a 3D model, and with a little bit of thinking and fiddling about it, you can get to know more about the structure and functionality of a thing. And then you can make it yourself [...] Yeah, and I guess, it's even the same with computers" (Jonas, 28 years; own translation). The fascination with the things that are created in FabLabs can be traced back to their being self-made, new and having a personal meaning. They express individuality and creativity and encourage the developers to continue crafting.

6. Conclusion

The fascination with being the maker, the fusion of materiality and virtuality, and reverse engineering were empirically analyzed on the basis of digitized objects that were produced in FabLabs during the workshop Shape your World. The theoretical frame consists of developments such as the Internet of Things and postmodern sciences, which deal with materiality and the objects. For the object analysis, object research has been introduced in order to create an approach towards the things and their inventors.

The English anthropologist Daniel Miller (2008) also dealt with the things and chose a highly interesting approach to get to know them. However, he does not solely concentrate on the digitized things, as I do in this paper. He rather analyzes the everyday things to be found in houses and apartments as part of our daily surroundings. In order to do so he chose a street in London and questioned hundred persons living there. His interest was to find out about the things in their homes in order to draw connections between the persons and the things. In his book The Comfort of Things, he portrays selected stories.

By contrast, I tried to get information on the things themselves through analyzing them. Based on that I investigate connections between humans and things. It has been discussed why self-made things attract so many, and what is new about the relation between materiality and virtuality. FabLabs provide the research laboratory for such analysis, where software and hardware are provided for the production of things. With the aid of software, 3D models are made on computers and are then transferred to fabrication devices in a particular format which then again construct a physical object of it. A 3D printer practically becomes the link between the immaterial/digital sphere and the material objects.

Some find it contradictory that our culture is becoming more and more digitized, the computer being part of everyday life, while at the same time the demand for material things is growing. Things exist for human beings in order to use them and discard them after usage. "Our everyday life is dominated by the existence of things" (Liessmann 2010, p. 11; own translation). Aspects such as the things' past or future cannot be recalled. We consume goods from all over the world without even knowing where they really come from. Likewise, what happens with those things we throw away remains a black box. For Liessmann (2010) the question where mankind comes from and what is going to happen after its existence has been expanded towards the question where the things come from and what is going to happen to them in the future. FabLabs provide the possibilities to recollect this. They are places where an object can be produced, from its first idea to its digitization to its materialization, and where the objects' biographies become transparent.

References

Ackermann, E 2007, 'Experiences of Artifacts: People's Appropriations / Objects 'Affordances'', in von Glasersfeld, E (ed.), *Keywords in radical constructivism*, Sense Publishers, Rotterdam, Taipai, pp. 249-259.

Anderson, C 2013, *Makers: The New Industrial Revolution*, Hanser Verlag, München.

Arendt, H [2010] 1960, *Vita activa oder Vom tätigen Leben*, Piper, München.

Boeing, N 2010, *The Future is Fab*, media release, 3 March, Heise Online, viewed 10 January 2013, <http://www.heise.de/tr/blog/artikel/The-future-is-fab- 942575.html>.

Breckner, R 2008, 'Bildwelten – Soziale Welten. Zur Interpretation von Bildern und Fotografien', *Online-Beitrag zu Workshop & Workshow* vom 23./24.11.2007, viewed 12 February 2013, <www.univie.ac.at/visuellesoziologie/>.

Büching, C, Walter-Herrmann, J & Schelhowe, H 2013, 'Lernen in Interaktion mit digitalen Medien', in Carstensen, T, Schachtner, C, Schelhowe, H & Beer, R (eds.), *Digitale Subjekte. Praktiken der Subjektivierung im Medienumbruch der Gegenwart*, transcript, Bielefeld.

Dourish, P 2004, *Where the Action Is. The Foundations of Embodied Interaction*, MIT Press, Cambridge, MA.

Frisch, M [2008] 1957, *Homo faber: ein Bericht*, Suhrkamp, Frankfurt/Main.

Gershenfeld, N 1999, *When things start to think*, Henry Holt and Co, NY.

Gershenfeld, N 2005, *Fab. The coming revolution on your desktop – from personal computers to personal fabrication*, Basic Books, NY.

Herobuilders 2012, viewed 7 September 2012, <www.herobuilders.com/custom/index.htm>.

Latour, B 2005, *Reassembling The Social*, University Press, Oxford.
Liessmann, KP 2010, *Das Universum der Dinge – Zur Ästhetik des Alltäglichen*, Paul Zsolnay Verlag, Wien.
Mattern, F 2004, 'Ubiquitous Computing: Schlaue Alltagsgegenstände – Die Vision von der Informatisierung des Alltags', Bulletin SEV/VSE, no. 19, pp. 9-13, viewed 12 February 2013, <http://vs.inf.ethz.ch/publ/papers/mattern2004_sev.pdf>.
Miller, D 2008, *The Comfort of Things*, Polity Press, Cambridge, UK.
Neef, A, Burmeister, K & Krempl, S 2005, *Vom Personal Computer zum Personal Fabricator – Points of Fab, Fabbing Society, Homo Fabber*, Murmann, Hamburg.
Rammert, W & Schulz-Schaeffer, I 2002 (eds.), *Können Maschinen handeln? Soziologische Beiträge zum Verhältnis von Mensch und Technik*, Campus, Frankfurt/Main.
Rammert, W 2007, *Technik – Handeln – Wissen. Zu einer pragmatischen Technik- und Sozialtheorie*, VS-Verlag, Wiesbaden.
Papert, S 1980, *Mindstorms: Children, computers, and powerful ideas*, Basic Books, NY.
Papert, S 1987, 'Constructionism: A New Opportunity for Elementary Science Education', Award Abstract, viewed 12 February 2013, <http://nsf.gov/awardsearch/showAward.do?AwardNumber=8751190>.
Piaget, J 1974, *Theorien und Methoden der modernen Erziehung*, Fischer, Frankfurt/Main.
Schelhowe, H 2007, *Technologie, Imagination und Lernen – Grundlagen für Bildungsprozesse mit Digitalen Medien*, Waxmann, Frankfurt/Main, NY, Münster.
Schmidt, H 2011, *Die programmierbare Welt kommt*, media release, 23 December, Frankfurter Allgemeine Zeitung, viewed 12 February 2013, <http://www.faz.net/aktuell/wirtschaft/digitalisierung-die-programmierbare-welt-kommt-11573198.html>.
Schütz, A 1974, *Der sinnhafte Aufbau der sozialen Welt. Eine Einleitung in die verstehende Soziologie*, Suhrkamp, Frankfurt/Main.
Sterling, B 2005, *Shaping Things*, MIT Press, Cambridge, MA.
Turkle, S 1984, *The second self: computers and the human spirit*, Simon and Schuster, NY.
Turkle, S (ed.) 2011, *Evocative Objects. Things We Think With*, MIT Press, Cambridge, MA.
Weiser, M 1991, 'The Computer for the Twenty-Fist Century', *Scientific American*, vol. 265, no. 3, pp. 66-75.

MAKER CULTURE

NOTES ON MAKER CULTURE

EVA-SOPHIE KATTERFELDT,
ANJA ZEISING,
MICHAEL LUND

Our names are Anja Zeising, Michael Lund and Eva-Sophie Katterfeldt. We work at the University of Bremen in the research group Digital Media in Education. Whereas I (Anja) have a background in Computer Science and my PhD research questions revolve around designing digital media and interactions beyond the desktop with and for children. I (Michael) have a background in Art History, Cultural Studies, Media, and Cultural Mediation. I (Eva-Sophie) have a background in Digital Media. For my PhD thesis, I investigate aspects of modeling (in terms of designing, developing, creating) in Computer Science, Interaction Design and maker practices, trying to find informal ways of modeling to support maker projects. We have been working together to research concepts, methods, and technologies to motivate children and teenagers to learn about Digital Media. The combination of the technological, engineering perspective and the artistic, cultural point of view seems to be promising, perhaps even crucial, in the research field of Interaction Design and children, as well as in the field of recent developments in the fields of both maker culture and FabLabs.

The maker culture movement was started decades ago. In the sixties and seventies, a socio-cultural movement took place in the UK. Based on a more participatory understanding of politics and administration, people aimed at more influence on their living conditions, especially housing. In the 1970s in the US, social movements influenced pop culture: Topics like global economy, production processes, use of materials, and cooperation in using technologies were discussed with a new ecological perspective. Moreover, in the field of media production and journalism, movements like those of independent filmmakers arose to build communities for sharing knowledge, technologies, and financial resources. In the field of education, Dewey (1949) has paved the way for maker culture within the last twenty years. His motto 'Learning by Doing' contributed to a different understanding of learning in schools.

For several years now, this new flourishing of the maker culture can be observed. This revival has been driven by technological and societal developments such as Web 2.0, the Open Source idea, discomfort with the consumer society, desire for individuality, sustainability and manual 'real' activity as compensation for 'virtual' office work. But not only static, handicraft artifacts are constructed. Technical possibilities provided by Web 2.0 and Open Source resources blend into traditional activities. Makers also design, program, construct and engineer technically enhanced artifacts. This group has been coined as makers, referencing to the Make: Magazine (Mota 2011). Makers act as amateur interaction designers, crafters and engineers, creating their own personally meaningful TUIs (Tangible User Interfaces), sharing their projects and supporting each other in Web 2.0 communities. The Internet offers a vast amount of easily accessible resources and assistance, which in an offline world would be much harder to access. It also opens up opportunities for people with a special interest, who may find no people with similar interests in their local area. Let's illustrate this with an example: In order to find inspiration, tutorials or hints to bring my current DIY project a step further, I only need to type my idea into the text field of a search engine. In case I am not familiar with, e.g., how to connect LEDs to my controller, I find a video and tutorials showing me how to do. If I need a specific material like conductive fabric (which the local haberdashery shop surely has not in stock), I find online shops where I can order it immediately. Whenever I have technical problems with the implementation of my project, I can browse forums or open itself a thread with my question, usually bringing up a solution more quickly than if I had tried for days on my own. Finally, I can proudly publish and share my final project on a maker platform like Instructables or (in case of a 3D model) Thingiverse, maybe along with a tutorial and advice from my personal experience.

Whereas in the past, DIY activities where often motivated by lack of capital or material resources (and still be in many areas of the world), the motivation today has partially changed. DIY and maker projects showcased in the Internet often reveal a very hip, aesthetical, designedly style. This maker or DIY movement often appears as a life-style, factors such as self-expression in a mass culture seem to play a role. Formerly private projects can get an audience, reward is not only limited to your family and friends. Gauntlett (2011) addresses the question why everyday creativity is important to society. In his opinion, people doing DIY engage actively with their environment and make contact to others by means of their work, as opposed to passive consumers. He reasons that craft activity is no longer dominated only by professionals, but many amateurs who share their ideas and works on the Internet – resulting in a shift from a "sit-back-and-be-told culture" (Gauntlett 2011, p. 245) to a "making-and-doing culture" (Gauntlett 2011, p. 245).

To participate in today's DIY movement, media literacy becomes crucial in order to find the right information or to verbalize appropriate questions or to produce a comprehensive tutorial. Furthermore, getting involved in making digital artifacts or 3D models requires at least basic engineering and programming skills. At the same time, handicraft skills become less important in order to create artifacts. FabLabs offer professional tools and personal support beyond the home context and web communities. By means of 3D printers, cutters and other machines provided, FabLabs professionalize crafting. In order to cut out a complex shape for example, sawing skills are no longer required. Creating physical things becomes accessible for everyone – everyone who knows how to use the digital tools! Thus, FabLabs open up new possibilities of product manufacturing, professional prototyping and engaging with technology. Since crafting by hand becomes obsolete and mechanized, the maker can concentrate on other aspects, maybe on making his/her product more interactive. By means of FabLabs, one can combine computing skills with crafting and making. Shaping happens digitally, fabber and crafter apply different skills, but have the same goal – a personal meaningful artifact. People from different fields of study, such as crafting, education, design, industrial design, engineering, computer science, art, and socio-culture are fascinated by the idea of FabLabs and contribute to it, hence building up a FabLab community. Our view on FabLabs highlights its correlation of technology, design, and social interactions.

Concerning social participation, FabLabs provide resources for empowerment, as fabbers shape and construct their own products. Design culture may be enhanced by new methods of planning and implementing projects, and new forms of doing research with and through FabLabs can emerge and be established. For example, FabLabs could be utilized in science, where design is constantly becoming more recognized, as a technique to understand and approach a research question. In all disciplines in which an object is invented and thus meant to encourage new approaches to research, fabbing can provide great potential, such as in archeology, architecture, and engineering. But we also discussed the integration of FabLab technologies in everyday life. Fabbing seems to be a current phenomenon, which needs to be discovered, and its main purpose and benefits shall be explored.

In the future, FabLab technology may be established in schools in different variations and contexts. For example, laser cutters may be integrated in crafting, training collaboration and self-efficacy. As we have seen, many people can now easily engage as interaction designers and engineers of tangible digital artifacts. However, when browsing social Web 2.0 communities, there seems to be a gap between women, who craft, and man, who make. In today's society, knowledge and skills about digital media are essential (e.g. to participate on the job market and in the professional design of IT). Apart from curricula, we should think about educational tools supporting diverse people in having motivating experiences while engaging in technical issues. For example, modeling tools could be developed which lower the barriers to construct and program tangible devices or to create 3D models, while at the same time making those processes more transparent for the creators. Furthermore, environmental aspects should be considered to be more important, and therefore the usage and re-usage of recyclable materials should be discussed. Outside of school, the FabLab idea may potentially induce diverse activities. Maybe this idea will encourage discussions of design as a research method. The developments in this field seem to change the repertoire of research methods as well. In Design, Industrial Design and even in the Social Sciences some researchers believe in the potential of design as a research technique. We would like to see people from several backgrounds, as we are, collaborating and creating innovations.

References

Dewey, J 1949, *Demokratie und Erziehung. Eine Einleitung in die philosophische Pädagogik*, Westermann, Braunschweig, Berlin, Hamburg.

Gauntlett, D 2011, *Making is Connecting*, Polity, Cambridge.

Mota, C 2011, 'The rise of personal fabrication', in *Proceedings of the 8th ACM conference on Creativity and cognition*, C&C '11, ACM, New York, pp. 279-288.

THE HISTORY OF PRODUCTION WITH COMPUTERS

BERNARD ROBBEN

1. The Factory Worker: Architect or Bee?

Studying the history of production with computers has to go beyond the notion of impacts that digital technologies have on society; it rather implies studying new "Digital Formations" (Latham & Sassen 2005, p. 8) generated by computer-based design and manufacturing practices. There are many forms of intersections of society and technology. Examining the specific ways in which technologies are embedded in the context of daily life requires analyzing the mediating cultures that organize the relation between these technologies and its users. Putting these reflections in a historical context, the following famous quote cannot be ignored: "A spider conducts operations that resemble those of a weaver, and a bee puts to shame many an architect in the construction of her cells. But what distinguishes the worst architect from the best of bees is this, that the architect raises his structure in imagination before he erects it in reality. At the end of every labour-process, we get a result that already existed in the imagination of the labourer at its commencement. He not only effects a change of form in the material on which he works, but he also realises a purpose of his own that gives the law to his modus operandi, and to which he must subordinate his will. And this subordination is no mere momentary act. Besides the exertion of the bodily organs, the process demands that, during the whole operation, the workman's will be steadily in consonance with his purpose" (Marx 1990, p. 124).

Using the 'architect or bee' comparison, Karl Marx analyzes the role of labor in the dawn of mankind. The Industrial Revolution in connection with the birth of capitalism changes the nature of human labor. Division of labor under capitalist relations of production leads to the alienation of the worker from the produced product. Capitalists appropriate the intellectual labor of the engineer and the industrial designer. New dialectics between intellectual and manual labor shape many cultural relations of the capitalist society (Sohn-Rethel 1977). In order to understand the dialectics of these interwoven processes, the social history of the machine which supersedes manual work must be reflected, as well as the development of the computerized media which substitute intellectual work. In our days, not only energetic machines but also symbolic machines determine the mode of production.

The introduction of symbolic machines – which are both a virtual and a real augmentation of mechanical machines – constitutes a new intertwinement of intellectual and manual labor. Their dialectics can be characterized by a computer's Janus faced biformity: on the one hand controlling real processes, and on the other hand representing them. Machine control is transformed into information processing, and representation systems are transformed into informational machines. The machine becomes a medium, the medium becomes a machine. Producing and designing objects is indissolubly interwoven.

The following sketch of the history of production with computers does not aim to describe a line of historical facts from the first introduction of computer numerical controlled machines to FabLab practices, but rather tries to write on the fabric of the aforementioned interrelated processes in order to deconstruct the myth of the machine and to undermine the 'architect or bee' comparison for human design activities.

2. The Myth of the Machine

"The human species' use of technology began with the conversion of natural resources into simple tools. The prehistorical discovery of the ability to control fire increased the available sources of food and the invention of the wheel helped humans in travelling in and controlling their environment" (Wikipedia, entry 'technology'). In this narrative tool, making appears as the dawn of mankind, and human history is told as a progressive evolution from the hand-axe to modern factories (Rauter 1977). The development of technology is of course intertwined with the formation of civilized societies, but the view of human beings as tool-making animals constricts human history to a story of craftsmen and warriors and misremembers the role of domestic care and female artisans. Lewis Mumford's (1966/64) oeuvre deconstructs the notion of the primal man as tool-making creature. Mumford could not accept the idea that the advantage man had over the rest of the animals was that he was able to smack rocks against each other to make axes or sharpen sticks for digging. Instead, Mumford emphasized the role of language. Language capacitates human beings to communicate with other members of the species and to create symbols to represent experiences, objects, and memories.

The word technology originates from the Greek term τέχνη (téchnē), meaning 'art, skill, craft'. τέχνη joins tool making and the making of art objects, and combines utilitarian activities with play. Johan Huizinga (2008) suggested that play is a fundamental condition for the formation of culture. The paleoanthropologist André Leroi-Gourhan (1993) developed a theory for the understanding of the relation between the awakening of human intelligence and technology. The bonding of tool making and using with the power of language gives rise to a powerful development denominated by Leroi-Gourhan as the externalization of the human memory, which becomes possible through the interconnection between the twin poles of hand/eye and ear/mouth. This interconnection constitutes the parallel development of spoken language and of mark making. "Thus graphism, the making of visual marks, which was aligned to the hand and eye and which are neither simply proto-writing nor naïve forms of figurative representation, and oral language, which was bound up with the mouth and ear, developed in parallel and lockstep, but in balance, with neither dominating the other" (Gere 2006, p. 17).

In history, the sphere of production has always been linked to the sphere of art, science, and technology. Early civilizations like the Egyptian, Babylonian, and Greek were concerned with the solution of practical problems such as counting, the construction of buildings, calendars, and geometric representations. Before the Industrial Revolution, manufacturing was carried out by skilled artisans. Training was by apprenticeship. For calculation purposes the abacus was widely used by craftsmen and merchants. The incipient age of mass production marked the beginning of a change. Calculation became arithmetic according to the methods of Adam Ries and had to be written down in account books. As one of the first commodities of mass production, the printed book arose. The automatic weaving loom simplified the process of manufacturing textiles at large scale. Jacquard's pattern weaving loom codified the actions of the human weaver and converted them into marks on cards that were 'read' by the machine in order to repeat and execute the weaving automatically. Charles Babbage (1832), a mathematician and theoretician of labor management, developed the concepts of the Difference Engine and the Analytical Engine as a basis for the economy of machinery and manufactures. The engines were modeled on industrial machinery. Ada Lovelace, who worked together with Babbage, remarked that the machine combined its numerical quantities "exactly as if they were letters [and weaved] algebraical patterns just as the Jacquard-loom" (Gere 2008, p. 28) weaved fabric.

Babbage's and Lovelace's concepts and inventions were ahead of the possibilities of real manufacturing machines of their times. But Jacquard`s loom and the Spinning Jenny – invented by James Hargreaves – pushed the mechanization of the textile industries. The technology which stored the fabric design in a series of punched cards marks the beginning of the Industrial Revolution and the transition from manual labor towards machine-based manufacturing.

The introduction of steam power and powered machinery led to a dramatic increase in production capacity. "The great economic revolutions in history occur when new communication technologies converge with new energy systems" (Rifkin 2011, p. 1). Iron, steam technologies, and textile production were the basis of the First Industrial Revolution. The Second Industrial Revolution, which revolved around steel, railroads, electricity, and chemicals, contributed to the enormous advance of mass production. The automobile industry, especially the Ford Motor Company that introduced the moving assembly line, can be considered a paradigmatic example; therefore, the related economic and social system has been called Fordism.

In the 19th century, the typewriter standardized and mechanized language writing, reducing its elements to a system of abstracted signs. Alan Turing translated this system into a universal machine concept. "The operations of capitalism are fundamentally predicated on abstraction, standardization and mechanization, to ensure that it can operate as universal machine, capable of treating disparate phenomena as equal and interchangeable" (Gere 2008, p. 24). Under the conditions of capitalism, goods are not valued for their material and embodied usefulness, but for their exchange value. Goods as well as labor become commodities that can be exchanged for money. "Turing's imaginary device not only invokes the typewriter, one of the paradigmatic information technologies of nineteenth-century capitalism, but also, in the tape and writing head assemblage, the very model of the assembly line. Moreover, the algorithmic method which his machine was intended to automate is itself a model of the division of labor" (Gere 2008, p. 25). Cheap computing power fosters new techniques of automated production, which "are bringing about a revolution in manufacturing greater than anything since the Industrial Revolution" (Forester 1987, p. 171). Jeremy Rifkin hopes that Internet technology and renewable energy are merging to create a powerful Third Industrial Revolution. "The Third Industrial Revolution is the last of the great Industrial Revolutions and will lay the foundational infrastructure for an emerging collaborative age. The forty-year build-out of the Third Industrial Revolution infrastructure will create hundred of thousands of new business and hundreds of millions of new jobs. Its completion will signal the end of a two-hundred-year commercial saga characterized by industrious thinking, entrepreneurial markets, and mass labor workforces and the beginning of a new era marked by collaborative behavior, social networks, and boutique professional and technical workforces" (Rifkin 2011, p. 5). But before I can discuss such ideas of visionary utopia, I need to describe the elements of the industrial production systems based on computers.

Figure 1:
Close-up view of the punch cards used by Jacquard loom on display at the Museum of Science and Industry in Manchester, England (Source: Photography by George H. Williams. Wikipedia; this file is licensed under the public domain).

3. CAD and CAM in Global Production

The automation of machine tools is based on a programmable logic controller (PLC) which controls both the manufacturing machines and the assembly line. PLCs are general-purpose programmable devices and are able to substitute thousands of relays, cam timers, drum sequencers, and dedicated closed-loop controllers. Computer Numerical Control (CNC) machine tools revolutionized the machining processes. Computers are used to assist in all operations of a manufacturing plant, including planning, management, transportation, and storage, transforming the traditional assembly line to Computer Aided Manufacturing (CAM). This is not only a concern of manufacturing machines but also of the design and engineering process for new products. Concepts of the digital factory include also management and administration, sales and stewardship commitment. Computer Aided Engineering (CAE) systems are considered a node on a total information network and each node may interact with other nodes in the network. Computer Aided Design (CAD) means the use of computer systems to assist in the creation, modification, analysis, or optimization of a design. The era of computer-based production started in the 1960s. An early commercial application of computer aided manufacturing was Pierre Bézier's work developing the CAD/CAM application UNISURF for car body design and tooling at Renault. Bézier Curves are completely contained in the convex hull of the control points, so that the points can be graphically displayed and used to manipulate the curve intuitively. Hence, they are an example of how new forms of production generate new types of modeling tools and vice versa. Sketchpad, a revolutionary computer program written by Ivan Sutherland in 1963, is considered the ancestor of modern Computer Aided Drafting programs. Ivan Sutherland (2003) demonstrated that computer graphics could be used for both artistic and technical purposes in addition to showing a novel method of human-computer interaction.

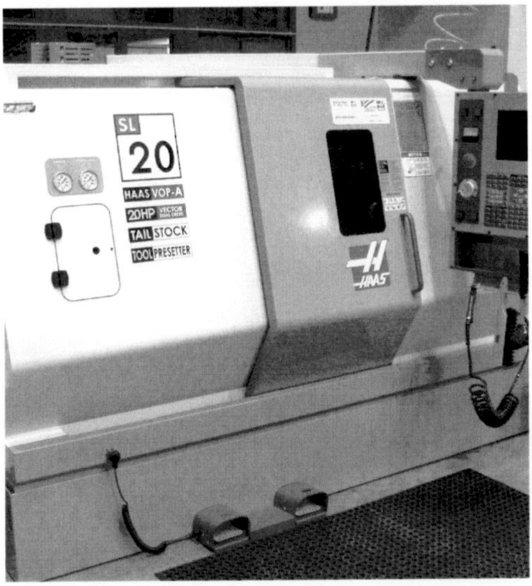

Figure 2:
Small CNC-engine
(Source: Photograph taken by Nathaniel C. Sheet. Wikipedia; this file is licensed under the Creative Commons Attribution 3.0 Unported license).

The traditional production process started with hand-drawings that were transferred into hand-made physical models out of wood or clay for visualization. Furthermore, technical drawings were developed on drawing boards by hand. The assembly line of mass production was rigid and inflexible. With the introduction of CAD and CAE/M, technical drawings became complex 3D models. Produced with CAD methods, 3D models facilitate complex simulations in multiple areas, for example, computer simulations of the production process and simulations of the functionality of the produced products. Engineers create computer simulations in virtual 3D space to explore the product from every angle and to zoom into details, and to model its material qualities for virtual testing. In the industry, the product lifecycle management (PLM) – the process of managing the entire lifecycle of a product from its conception through design and manufacture to service and disposal – is simulated in models of digital fabrication. In modern factories, welding, painting, assembly, pick and place, product inspection, and testing are typical applications for industrial robots. The composition of CAD, CAE/CAM in combination with robotic cells allows for the organization of the whole production process as a flexible manufacturing system.

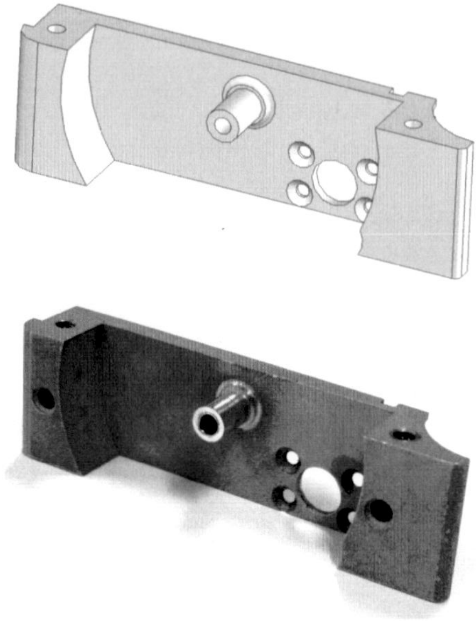

Figure 3: Model and processed part (Source: Wikipedia; this file is licensed under the public domain).

4. The Vision of Computer Integrated Manufacturing

The specter of capitalism dreams of uniting all of these computer-aided practices in one system. Computer Integrated Manufacturing (CIM) is the approach to control the entire production process with computers. Visions of lights-out manufacturing came up in the 1970s. Would the transformation of the manufacturing process lead to factories that are fully automated and require no human presence on-site where raw materials enter and finished products leave? Utopian views that robots will do all the work and human beings can spend their whole life in leisure seemed to be realistic – for some as a horror vision where the machines take command, for others as the promise of a land of wealth and freedom.

On the one hand, there was the widespread belief that the computerization of the manufacturing process would free human beings from soul-destroying, backbreaking labor and leave them free to engage in creative work. On the other hand, empirical studies about the impact of the computerization on the new relation between manual and intellectual work showed negative or problematic consequences for job satisfaction: "As the medium of knowing was transformed by computerization, the placid unity of experience and knowledge was disturbed. Accomplishing work depended upon the ability to manipulate symbolic, electronically presented data. Instead of using their bodies as instruments of 'acting-on' equipment and materials, the task relationship became mediated by the information system. Operators had to work through the medium of what I will call the 'data interface', represented most visibly by the computer terminals they monitored from central control rooms. The workers in this transition were at first overwhelmed with the feeling that they could no longer see or touch their work, as if it has been made both invisible and intangible by computer mediation" (Zuboff 1988, p. 62; his emphasis).

Will the computer reduce us to robots? Or will it free our creativity and liberate us from drudgery? To escape from such simplistic alternatives, it is still useful today to read the study about the new human-machine interaction of Mike Cooley, an activist, engineer and academic of the 1970s and 1980s. Cooley deconstructed the often-stated opposites of the precise, reliable and fast machine and the slow, inconsistent, but highly creative human being. "Design methodology is not such that it can be combined at some particular point like a chemical compound" (Cooley 1987, p. 38). Producing a new 'whole' is a complex and extremely subtle area. CAD systems may lead to work which is alienating and of inhuman tempo. "In Britain and much of Europe up to the 1940s the draughtsman was the center of the design activity. He could design a component, draw it, stress it out, specify the material for it and the lubrication required. Nowadays, each of these is fragmented down to isolated functions. The designer designs, the draughtsman draws, the metallurgist specifies the material, the stress analyses the structure and the tribologist specifies the lubrication" (Cooley 1987, p. 17). Workers lose their professional skills, for example, a welder who takes a robotic welding device through the welding procedure of a car body is building skill into the machine and deskilling him or herself. The accumulation of welding experience is absorbed by the robot's self-programming systems.

Cooley formulated a radical critique of such practices, which conceptualize western science and technology as efficient and neutral. He associated its practices to fragment labor to mindless functions with the problem that they have the "historically determined male values built into them. These are the values of the white male warrior" (Cooley 1987, p. 17). In this time, Mike Cooley worked at Lucas Aerospace, a famous British manufacturer of components for the motor industry and aerospace industry, which was based in Birmingham. Nearly half of its business was related to military matters – in production of combat aircraft and the Sting Ray missile system for the NATO. In 1976, the militant workforce within Lucas Aerospace was facing significant layoffs. Under the leadership of Mike Cooley, they developed the Lucas Plan, which aimed to convert the company from an arms production factory to a manufacture of socially useful products, thus also aiming to save jobs. As a consequence, Cooley lost his job, and the plan was never put into place. But the ideas of Cooley and the struggle of the workers of Lucas Aerospace influenced the disputes about how human beings can become architects of a valuable production system and not unconscious bees that spawn damaging commodities.

Looking at it from today's perspective, CIM generated some automatic factories, which produce their products with little or no human intervention, like the Japanese robotics company FANUC or some branches of the automobile industry. But the main effect of computerization was the beginning of lean production and just-in-time production. Essentially, lean production is centered on preserving value with less work. This management philosophy is derived mostly from the Toyota Production System, which implements flexible mass production. Lean manufacturing claims to get the right things to the right place in the right quantity at the right time to achieve perfect work flow, while also claiming to be flexible and able to change. This includes the just-in-time production strategy that strives to improve a business return on investment by reducing in-process inventory and associated carrying costs. To meet these objectives, the process relies on constant flow of processed data between different points in the process, which tell the production machines when to make the next part. The acceleration of product cycles requires not only highly elaborated models of new products, but also hand-networked information systems to control the whole cycle, from the usage of the raw material up to the delivering of the product to the costumer.

For the system of production with computers, the shift from assembly-line work to automated production by robots is less important than the information technology's enormous potential to integrate global networks of distributed production and management systems. We are not living in the era of robots but in the era of ubiquitous pervasive computing, which has to be inscribed into the social and environmental complexity of our existing physical environment. The involved flexible mass production relies on a close coupling between virtual models of production and the real production process itself. Virtual models simulate the production process in a way that the simulated process is not only calculated in an abstract model but also visualized through a computerized artificial environment that can be actually experienced: Engineers can interact with their models. And the program code written for the control of the virtual model can be used to control the real CNC machines. In practice, the code still has to be refined and appropriated to the real circumstances. But manufacturing in the digital age relies on a reduplication of the involved processes in interactive computer models. In almost all branches of production, the importance of time and effort for the final real manufacturing decreases and the costs for design and management increase.

"What is important in all this is not only that the fragmented functions of the designer have been built into the computer, but that the highly skilled and satisfying work on the shop floor has been destroyed" (Cooley 1987, p. 17). Mike Cooley's analysis of the situation is at the same time absolutely true and somehow obsolete. As Daniel Bell (1973) predicted, the new division of labor results in a shift from manufacturing to services. The First Industrial Revolution superseded agricultural labor; the Third Industrial Revolution drives manual workers away. Production of material products needs less blue-collar employees. New science-based industries emerge, which need new technical elites. The digital network is the sphere where the spatial and temporal globalization of labor is made possible. Global labor is the endless recombination of a myriad of fragments that produce, elaborate, distribute, and decode signs and informal units of all sorts. Labor has become the cellular activity where the network activates an endless recombination.

5. Rapid Digital Manufacturing

Film industry first showed us digital manufacturing through the illusiveness of movies from Metropolis to Matrix. Today, 3D computer models can be presented as animated scene elements. In many respects the underlying geometric models are the same as in 3D models for manufacturing purposes. In both cases, they must be precise representations of real objects. But there are significant differences. Most important in the movie domain is the photo-realistic rendering, while in the manufacturing domain, the precise measuring and the representation of material qualities of the object are much more required. Therefore, 3D models for CAD do not only have to realize surface models but also solid models. Virtual drilling in movies or computer games only has to simulate the resulted whole, but in the context of design for manufacturing, it also needs to simulate the exact decrease of weight of the component part because the whole drilling must be captured. The simulation of pieces of work must include the various potentials to experiment its inherent qualities – for example, if the structure of the component is sufficiently stable. And it is of crucial importance that the digital model of such an element is embedded in models of manufacturing of the whole product and of distributing the related commodities. Digital prototyping gives conceptual design, engineering, manufacturing, and sales and marketing departments the ability to virtually explore a complete product before it is built. It enables product designers to operate moving parts, to determine whether or not the product will fail, and to see how the various product components interact with subsystems – either pneumatic or electric. Digital prototyping accelerates the product life cycle management.

Digital prototypes as interactively manipulable models have many advantages. But for some purposes physical prototypes are preferred to virtual ones, in particular because they are tangible. Advances in CAD technologies in combination with advances in material research fostered new rapid prototype techniques to quickly fabricate a scale model of a physical part or assembly.

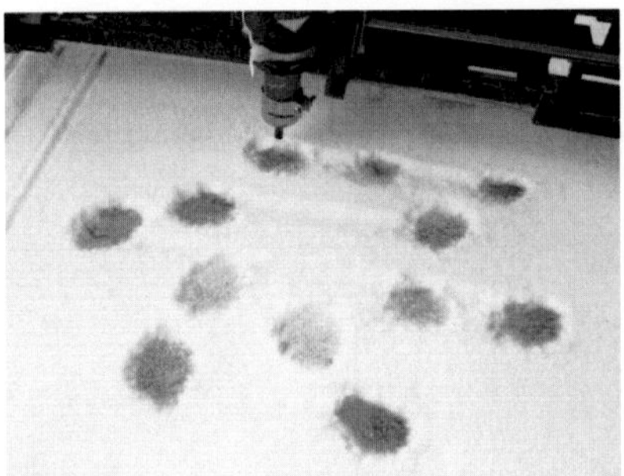

Figure 4:
An experimental rapid prototyping machine that can use granulated sugar as the printing medium (Source: Wikipedia; this file is licensed under the Creative Commons Attribution 3.0 Unported license).

"The main benefit to be gained by taking additive manufacturing approach [...] is the ability to manufacture parts of virtually any complexity of geometry entirely without the need for tooling" (Hopkinson et al. 2006, p. 5). Rapid manufacturing techniques allow to manufacture almost any shape that designers come up with, because they are no longer constrained by the necessity to produce parts in molds. New and exotic materials become available to manufacturing processes. The elimination of tooling opens up a pool of possibilities for low and medium volume manufacture. Advantages of additive manufacture over more conventional subtractive or formative methods come from these new possibilities of design. Without the cost of tooling to amortize into the parts produced, each component can be different, potentially allowing for true mass customization of every product.

Material research is one of the main stumbling blocks to the adoption of additive manufacturing. To explore the links between the processing of materials, the development of their microstructures, and their useful functional properties across a range of applications, and to seek to enhance the performance of existing materials and develop new materials with new functionalities constitutes one of the big challenges in this domain. New technologies for rapid manufacturing processes, such as stereolithography, jetting systems, direct light processing technologies, laser sintering, laser melting, and so on emerge (Hopkinson et al. 2006, pp. 55-80).

In the production domain, the merging of advanced CAD methods and additive manufacturing processes constitutes the emerging field of rapid manufacturing (Hopkinson et al. 2006). Rapid manufacturing utilizes CAD-based automated additive manufacturing to construct parts that are used directly as finished products or components. In additive manufacturing, 3D solid objects are produced from a digital model. An object is created by laying down successive layers of material. Traditional machining techniques rely on the removal of material by methods such as cutting and drilling. The use of additive manufacturing takes virtual designs from CAD or Animation modeling software transforms them into thin, virtual, horizontal cross-sections and then creates successive layers until the model is completed. Rapid manufacturing offers the potential of efficiently translating product innovations into prototypes and small production batches.

The first techniques for rapid prototyping became available in the late 1980s. Today, they are used for a much wider range of applications. Some sculptors use the technology to produce exhibitions. Intertwinement of CAD and rapid prototyping techniques with virtual modeling influenced new design thinking. This approach appreciates possibilities and opportunities more than realistic codification of given facts. In his novel The Man without Qualities, Austrian author Robert Musil (1997, p. 16) describes this way of thinking and interacting with the world as follows: "Someone who possesses this sense of possibility does not say for example: here this or that has happened, or it will happen or it must happen. Rather he invents: here this could or should happen. And if anybody explains to him that it is as it is, the he thinks: well, it probably could be otherwise." Part of the designers following such a design philosophy gathered in the Open Design movement that demands the free distribution and documentation of design, similar to the Open Source movement in the software domain (van Abel et al. 2011).

6. The FabLab Worker: Architect and Bee?

Neil Gershenfeld, who broadly explored how the content of information relates to its physical representation and how a community can be powered by technology at the grassroots level, founded the globally growing FabLab movement. He draws the vision of a revolution in manufacturing which leads from personal computers to personal fabrication. The FabLab idea is based on FabLab workshops which are generally equipped with an array of flexible computer-controlled tools that cover several different length scales and various materials like laser cutters, 3D printers and so on. But the global community, which runs websites like Thingiverse, dedicated to the sharing of user-created digital design files, is as important as the physical workshop.

Based on this new relationship of design and production, a new DIY culture has arisen, which is closely connected to the aforementioned Open Design movement. The active members of this movement cooperate in open community labs, incorporating elements of machine shops, workshops and/or studios where the activists come together to share resources and knowledge to build and make things. They appropriate inexpensive rapid manufacturing technology by constructing 3D printers which are able to print out new 3D printers. 'Do it yourself' becomes a slogan within a high-tech domain. High-tech devices become graspable again.

Many of these activists like to download some prefabricated designs from Thingiverse and just print it out. But the DIY culture in the creative environments of FabLabs experiments with new forms of design and manufacturing on a personal level. Are these activists the conscious architects of their products? Or do they act like inferior animals, which unconsciously build useless things? Perhaps, a more realistic view would be that they are both: sometimes taking up the role of an architect, sometimes the role of the bee. Creativity and innovation happen only when nurtured by large areas of common knowledge, which help to actively construct new things.

Putting such activities in a historical perspective fosters a reflexive and reflective attitude. Evaluating the potential of FabLab techniques "brings us to the question we must ask of every device – does it serve our human purposes? – a question that causes us to consider what these are" (Turkle 2008, p. 29).

References

van Abel, B, Evers, L, Klaassen, R & Troxler, P 2011, *Open design now – why design cannot remain exclusive,* BIS Publishers, Amsterdam.

Babbage, C 1846, *On the economy of machinery and manufactures,* J. Murray, London.

Bahr, HD 1983, *Über den Umgang mit Maschinen,* Konkursbuchverlag, Tübingen.

Bell, D 1976, *The cultural contradictions of capitalism,* Basic Books, New York.

Bell, D 1973, *The coming of post-industrial society: a venture in social forecasting,* Basic Books, New York.

Berardi, F 2009, *The soul at work – from alienation to autonomy,* MIT Press, Los Angeles.

Bessant, J & Chisholm, A 1989, *Human factors in computer-integrated manufacturing,* in Forester, J (ed.), *Computers and the Human Context,* MIT Press, Cambridge, MA.

Biswas, S & Lovell, BC (eds.) 2008, *Bézier and splines in image processing and machine vision,* Springer, London.

Cooley, M 1987, *Architect or bee? The human/technology relationship,* The Hogarth Press, London.

Forester, T 1987, *High-tech society – the story of the information technology revolution,* Blackwell, Oxford.

Galloway, A 2004, *Protocol – how control exists after decentralization,* MIT Press, Cambridge, MA.

Gere, C 2008, *Digital Culture,* Reaction Books, London.

Gere, C 2006, *Time art and technology,* Berg, Oxford, London.

Gosh, RA 2005, *CODE – collaborative ownership*, MIT Press, Cambridge, MA.
Hopkinson, N, Hague, R & Dickens, P (eds.) 2006, *Rapid manufacturing – an industrial revolution for the digital age*, John Wiley & Sons, Chichester.
Huizinga, J 2008, *Homo Ludens – the study of the play element in culture*, Routledge, London.
Latham, R & Sassen, S 2005, *Digital formations – IT and new architectures in the global realm*, Princeton University Press, Princeton, Oxford.
Leroi-Gourhan, A 1993, *Gesture and speech*, MIT Press, Cambridge, MA, London.
Marx, K [1990] 1867, *Capital*, vol. I, Penguin Books, London.
McCullough, M 2004, *Digital ground – architecture, pevasive computing, and environmental knowing*, MIT Press, Cambridge, MA.
Mumford, L 1966/64, *The myth of the machine*, Harcourt Brace Jovanovich, New York.
Musil, R 1997, *The man without qualities*, Picador, London.
Rauter, EA 1977, *Vom Faustkeil zur Fabrik – Warum Werkzeuge die Menschen und die Menschen die Werkzeuge verändern*, Weismann, Munich.
Rifkin, J 2011, *The third industrial revolution*, Palgrave MacMillan, New York.
Sohn-Rethel, A 1977, *Intellectual and manual labour: a critique of epistemology*, Humanities Press, Atlantic Highlands, NJ.
Sutherland, I 2003, *Sketchpad: a man-machine graphical communication system*, University of Cambridge Technical Report No. 574 based on a dissertation submitted January 1963 by the author, viewed 22 November 2012, <http://www.cl.cam.ac.uk/techreports/UCAM-CL-TR-574.pdf>.
Turkle, S (ed.) 2008, *The inner history of devices*, MIT Press, Cambridge, MA, London.
Zuboff, S 1988, *In the age of the smart machine – the future of work and power*, Basic Books, New York.

MAKER CULTURE, DIGITAL TOOLS AND EXPLORATION SUPPORT FOR FABLABS

EVA-SOPHIE KATTERFELDT

1. Introduction

FabLabs are situated in a so-called 'maker culture'. What this maker movement distinguishes from previous DIY cultures is the involvement of digital media in the creation process. FabLabs offer makers a wide range of opportunities to realize their projects. But in order to explore the whole range of making activities provided by FabLabs, one needs to know more about the underlying principles of digital models. The tools currently available seem either to be black-boxing the underlying principles, or are entirely targeting experts. So the question is, how can tools be designed to allow for a wider exploration?

I will first introduce the idea of maker culture and what distinguishes it from other DIY cultures. Who are the people making, how are they doing things themselves, what is their motivation, and what are their values? Further, I will show different levels of how digital media are inevitably involved in people's making, and will outline the current state of digital crafting tools. Finally, it will be concluded how these tools could be designed differently to expand what and how makers can do with digital media in FabLabs, in order to explore the environmental opportunities provided there in more detail.

2. Maker Culture

For almost a decade now the upcoming of a maker culture as a subculture of DIY culture has been propagated. What is so new about this phenomenon, who are these makers and what are they making?

To get an idea of what people are doing by themselves in context of the maker culture, the website Instructables[1] provides a good showcase. Instructables is a community platform where all kinds of makers present, share, and comment on their makings and the making of their projects by ways of step-by-step instructions illustrated with pictures and explaining text. The artifacts range from experimental

1 | For more information see the website, last viewed 7 January 2013 <http://www.instructables.com>.

cooking recipes to fancy knitting projects to home improvement to crazy toys and other gadgets. At the same time, people like to shop at Dawanda[2] or Etsy[3], where people sell their self-designed and manufactured artifacts. Maker fairs in the U.S. attract more than a hundred thousand people.

People have always been designing and creating things by themselves. Therefore, the idea of a DIY culture is not a new phenomenon. As examples for DIY cultures, people often refer to the subcultures situated in the punk, hippie or hacker movements in the U.S. and U.K. in the 20th century (Kuznetsov & Paulos 2010; Anderson 2012), whose DIY activities included not only crafting things, but also production and publishing of media such as magazines and music. They were led by political and social-ethical values, especially criticizing consumerism. Other political and societal forces motivating DIY cultures include questioning the positioning of the consumer society, like a desire for individuality, sustainability, and to do something 'real' as compensation for the predominant virtual office work (Anderson 2012, p. 13).

The term maker culture encompasses all kinds of DIY activities ranging from traditional crafting to engineering. However, what lets "enthusiasts", "hackers", "modders", "hobbyists" (Torrey & McDonald 2007), "amateurs" (Frauenfelder, 2010; Paulos, 2012) become coined as "makers" (Anderson 2012, pp. 20-21) is the fact that digital media are the pivotal element in their making. In most of the maker projects, digital media serve as communication tools for information exchange, community building and showcasing of DIY projects. These are web community platforms where documentations of created artifacts and information and skills concerned about how to make things are shared. More importantly, digital media also serve as tools for digital crafting, e.g. by using software to create printable 3D models. Moreover, digital media are even the target artifacts of some makers' projects, e.g. where makers enhance physical material with computational 'intelligence' (also known as physical computing). Self-made 3D printing machines like the RepRap[4] are examples of such advance.

People have enjoyed doing things themselves before any maker movement. But the Web 2.0 and other technical possibilities render them visible and allow DIYers to connect with each other regardless of one's actual location. Accordingly, Anderson (2012, p. 21) sees makers in the context of the Web generation to whom it is natural to share ideas and thoughts online. Through these new ways of sharing and through ideas such as Open Source licensing, people do not only get inspired by others but they can also take up the solutions of others and build on them, so that new ways of making on more sophisticated levels become possible. It is quite easy to find instructions for unconventional, creative projects on the web, and one can take up these ideas and develop them further.

2 | For more information see the website, last viewed 7 January 2013 <http://www.dawanda.de>.

3 | For more information see the website, last viewed 7 January 2013 <http://www.etsy.com>.

4 | For more information see the website, last viewed 7 January 2013 <http://www.reprap.org>.

It cannot be assumed that there are more people practicing DIY today because of the maker movement. The maker phenomenon is rather a new form of a DIY community whose making is inevitably influenced and linked to digital media – by ways of communication or design tools as well as by actually making their entirely digital or digitally enhanced material artifacts.

2.1 Makers, Motivation, Values

Many makers and their products do not correspond to the image of the hobbyist lingering at the home improvement store or the knitting granny; instead, they appear to be young, 'hip' and creative people with an attitude for design and arts. Many of the creations presented on the Internet reveal a modern design style and are not restricted to their functionality.

Mark Frauenfelder, editor in chief of Make: Magazine[5], gives insights into the maker scene and about the personal motivation that drives the people around him to make. Among the factors he mentions are challenging oneself, learning new skills, intellectual curiosity, relationship to objects, doing something with your hands instead of only virtually, tactile and haptic experiences, being creative, a desire for self-expression and acting out creativity and independence, sense of control, relaxing and meditation, creating things one cannot buy, or to personalize objects (Frauenfelder 2010, pp. 219-220).

According to a survey among 2600 members of online DIY communities, the main motivations to contribute to DIY communities are related to sharing and learning: "Motivations for contributing to DIY communities highlight information exchange as a core value: receiving feedback on projects, educating others, and showcasing personal ideas and skills are the top factors" (Kuznetsov & Paulos 2010, p. 302). Among those, to "get inspiration", to "learn new concepts" (Kuznetsov & Paulos 2010, p. 299), and to meet like-minded people were named as the top motivations to contribute (Kuznetsov & Paulos 2010). As a whole, participants emphasized "values of open sharing, learning, and creativity over profit and social capital" (Kuznetsov & Paulos 2010, p. 295).

Anderson (2012) accounts "a cultural norm to share those designs and collaborate with others in online communities" (Anderson 2012, p. 21) as another key characteristic of the maker movement, besides the use of digital tools. Making does not only happen at home or in virtual spaces, but locations for makers are also hacker spaces or FabLabs, who share those maker values. For example, one note on the Fab Charter says: "Fab labs are available as a community resource, offering open access for individuals as well as scheduled access for programs" (Fab Charter 2012).

3. Digital Media in a Maker Culture

Using and making digital media is an essential practice in the current DIY phenomenon, where people appear as makers. I distinguish here between three ways of how makers may be involved with digital media: Firstly, makers

5 | For more information see the website, last viewed 7 January 2013 <http://makezine.com>.

who consume, author, share their making by means of communicating in web communities. Secondly, makers who create their artifact by means of digital media as crafting tools, and thirdly, makers whose target artifact is an interactive product, a digital medium.[6]

3.1 Makers on the Web

"The growing interest in DIY is charging a virtuous circle – individuals who make things enjoy documenting their projects online, which inspires others to try making them, too" (Frauenfelder 2010, p. 222). As for everyone else, the Internet offers a wealth of quickly accessible, free resources and assistance to makers that is not so easy to access offline. It also opens up opportunities for exchanging, sharing and collaborating with others interested in a specific topic, for instance, those who can find little common ground in their local area.

On platforms like Instructables and Makezine[7], makers share their project documentations as tutorials. These so-called 'HowTos' are step-by-step instructions, which essentially contain images and little text, sometimes with videos or other attachments (e.g., program code). Examples of web platforms where mainly 'analog' projects concerned with home decoration, food and furniture are shared at Craftzine[8], Ravelry[9] for knitting and crocheting or Ikeahackers[10] for furniture modding. Platforms like Etsy[11], Cargoh[12] or Dawanda[13] open up opportunities to sell handmade artifacts, and also provide showcases to get inspiration for one's own projects.

For physical production, as it is provided by the machines at FabLabs, platforms such as Thingiverse[14], Ponoko[15], Studio:Ludens[16] have emerged. They offer not only communities but also large databases with 3D models and shapes ready to

6 | If one creates a digitally enhanced artifact, the other ways of using digital media as communication tool (browsing the web etc.) and as digital tools (e.g., writing the program code) are likely to be involved, too.

7 | For more information see the website, last viewed 7 January 2013 <http://makezine.com>.

8 | For more information see the website, last viewed 7 January 2013 <http://blog.makezine.com/craftzine/>.

9 | For more information see the website, last viewed 7 January 2013 <http://ravelry.com>.

10 | For more information see the website, last viewed 7 January 2013 <http://ikeahackers.net>.

11 | For more information see the website, last viewed 7 January 2013 <http://www.etsy.com>.

12 | For more information see the website, last viewed 7 January 2013 <http://www.cargoh.com>.

13 | For more information see the website, last viewed 7 January 2013 <http://www.dawanda.de>.

14 | For more information see the website, last viewed 7 January 2013 <http://http://www.thingiverse.com>.

15 | For more information see the website, last viewed 7 January 2013 <https://www.ponoko.com>.

16 | For more information see the website, last viewed 7 January 2013 <http://www.studioludens.com>.

be printed, or tools for creating or modifying them and even printing services, addressing experienced designers as well as unskilled amateurs. Besides those platforms targeting crafters and makers, other Web 2.0 forms like blogging or Youtube videos are used for sharing, as well as websites focusing on certain technologies, e.g. Arduino[17], and they all serve as knowledge base and provide information exchange by means of forums.

3.2 Making Digital Media

Makers who create technologically enhanced artifacts – digital media – e.g. by means of Open Source hardware, become amateur interaction designers. Many of these makers have no professional computer science or design education, but create computational artifacts rather informally, although they are based on formal computational models and concepts.

The technological development has not only contributed to the maker culture in the form of Web 2.0. PCB (Printed Circuit Board) computing technologies have also become accessible for people not skilled as engineers or programmers in recent years. The Open Source movement has contributed to this. One prominent example is the Arduino, an Open Source microcontroller board with a big community and many modifications around. They include variations like the LilyPad Arduino (Buechley et al. 2008), which relates to non-engineering practices like sewing and fashion and therefore shall attract more female and artistic makers than the other boards (Buechley & Hill 2010). With Arduino, people create all kind of physical computing artifacts. Within the context of FabLab machines, Open Source technology also plays an important role.

3.3 Digital Media as Crafting Tools

Besides being tools for networking or targets of one's making, digital media play an important role in many maker activities. Not only for programming hardware or composing media, but especially in FabLab contexts, digital media as crafting tools are an essential part of the making process. With FabLabs, new ways of manufacturing come to the people. On the one hand, some machines may make certain arts and crafts skills obsolete. On the other hand, it opens up new ways of production to the individual, opening up novel opportunities for DIY. But in order to create artifacts, one has to know how to create digital models. Certain software tools are offered to help doing this. Working with such software is either quite complex and requires a high mathematical understanding of geometry and computer graphics, or it is rather simple, like building with Lego™ bricks.

4. Digital Tools for Exploring FabLabs?

Values of maker culture and FabLabs, such as participation for everyone, have been mentioned above. One can say that these values correspond to the idea of "low threshold", "high ceiling" and "wide walls" (Resnick et al. 2005, p. 5) as

17 | For more information see the website, last viewed 7 January 2013 <http://arduino.cc>.

requirements for tools that support creativity and designing.[18] Tools or construction kits have a high ceiling if they are powerful enough to allow for creating comprehensive, sophisticated projects. Wide walls means that they support a diversity of projects of different types, inviting makers to explore the design space.

I want to pick up the point made by Heidi Schelhowe (see Schelhowe in this book) that FabLabs offer a low threshold, arguing that the general idea of FabLabs offers not only a low threshold, but also wide walls and high ceiling. To illustrate this with an example: One can start by downloading a digital model from the web, then open the file with a tool to print it, or send it over to a 3D printing organization or someone at the FabLab who can print it. This is a low threshold. But one could then continue using digital crafting tools, like 3D programs, to customize that model one wants to print, to create new models, and so on, hence exploring a range of possibilities like moving along on wide walls. At a later point, one might wish to improve the printing output and thus get involved in the printing process, and eventually assemble or even develop one's own 3D printer, and thereby – literally speaking – climb up the ceiling.

Thus, FabLabs can provide experiences on certain levels where digital media are involved, and thus open up wide walls and high ceiling to the level of engagement. The particular technologies available in FabLabs, as well as the unlimited possibilities what to create with them, allow for wide walls. Creating physical objects requires special thinking and an understanding of computational 3D modeling. Furthermore, creating interactive, digitally enhanced artifacts is complex, involving programing, electrical engineering and interaction design. The values of maker culture and FabLabs count on mutual sharing and learning through peers. However, not always can a peer be around in the construction process, so we should consider how to incorporate those values into digital tools involved in making.

I argue that in order to be able to explore the possibilities FabLab equipment can offer, it is necessary that one can get an understanding of the fundamental principles of the digital models, i.e. to make the hidden numbers and vectors more transparent and adjustable. This should let us consider the tools involved in making.

One statement of the Crafter Manifesto claims: "Essential for crafting are tools that are accessible, portable, and easy to learn" (Muutanen 2005, p. 7). These aspects are definitely important, but do not consider how to support makers who gradually want to advance further in exploring the possibilities of making. Useful guidelines for designing and developing platforms that support maker activities in the creation of digitally enhanced artifacts can also be found in the "manifesto for diy internet-of-things creation" (De Roeck et al. 2012, p. 174). The statements of the manifesto relate to maker practices of sharing and creativity, and also consider different levels of computational skills to support everyone, from amateur to professional. For instance, the second statement claims to "support a spectrum of expertise in computational thinking by offering different layers of computational abstractions" (De Roeck et al. 2012, p. 173).

18 | Resnick and Silverman (2005) and Resnick et al. (2005) give these requirements for designing construction kits for children as well as general "design principles for tools to support creativity thinking" (Resnick et al. 2005, p. 3).

I argue that not only different levels of abstractions should be supported, but also the switching and moving between levels when doing activities like creating 3D models or programming digitally enhanced artifacts (for an approach see Katterfeldt & Schelhowe 2008). Among the Arduino community, graphical programming languages like Modkit (Millner & Baafi 2011) or Amici (Dittert et al. 2012) have been developed. They allow creating programs by means of aligning graphical blocks, which at the same time generates program code. The code is not hidden, so that the user can observe the relation between the blocks and the written code and eventually get involved with coding. Tools like Fritzing[19] offer similar concepts for hardware creation, allowing to switch between iconic representations of breadboards and physical parts to the symbolic representation of the circuit, and eventually create a file for a PCB milling machine, which some FabLabs provide.

For creating printable 3D models, certain tools are available. Tools requiring relatively little knowledge about 3D modeling are Google Sketch Up[20], 123dapp[21] or Tinkercad[22]. But if one wants to get more control over the virtual objects and to start manipulating the underlying vectors and numbers, professional software tools like Solidworks[23] or Autocad[24] are more suitable, but require a higher degree of computational thinking. Tools supporting certain levels of knowledge and gradual levels of transparency are still a lack.

Furthermore, to acknowledge makers' ways of sharing and learning from peers, tools that integrate communication 'and' crafting might be beneficial, too. This could offer new possibilities on the meta level – the design process itself – by means of sharing undone project ideas or problems appearing during the process, guiding the process and the different issues involved, as well as documenting both successful and misleading ways of approaching complex issues.

5. Conclusion

In this chapter, I have introduced the maker culture, its values, makers' motivations and what is special about it – the involvement of digital media on several levels. I have drawn distinctions from the maker cultures' values and the idea of FabLabs to the concepts of low threshold, wide walls and high ceiling and have argued that FabLabs can very well fulfill these requirements. I have pointed out that there is

19 | For more information see the website, last viewed 7 January 2013 <http://fritzing.org>.

20 | For more information see the website, last viewed 7 January 2013 <http://www.sketchup.com/intl/en/product/gsu.html>.

21 | For more information see the website, last viewed 7 January 2013 <http://www.123dapp.com>.

22 | For more information see the website, last viewed 7 January 2013 <https://tinkercad.com>.

23 | For more information see the website, last viewed 7 January 2013 <http://www.solidworks.com>.

24 | For more information see the website, last viewed 7 January 2013 <http://usa.autodesk.com/autocad/>.

still a lack of tools that allow us to move from one level of involvement with digital media to the next (and maybe back), and have given implications for designing tools that support digital crafting and at the same time allow for a better understanding of the computational concepts behind it. For the future, I hope that we will see more tools that allow for moving up the ceiling as well as exploring wide walls in FabLabs and among maker communities in general. This could encourage more makers not only to appear on the web, but also to move on to other forms of involving – to crafting with and creating of – digital media.

References

Anderson, C 2012, *Makers: The New Industrial Revolution*, Crown Business, New York, USA.

Buechley, L & Hill, BM 2010, 'LilyPad in the wild: how hardware's long tail is supporting new engineering and design communities', *Proceedings of the 8th ACM conference on designing interactive systems*, pp. 199-207.

Buechley, L , Eisenberg, M, Catchen, J & Crockett, A 2008, 'The LilyPad Arduino: using computational textiles to investigate engagement, aesthetics, and diversity in computer science education', *CHI '08: Proceeding of the twenty-sixth annual SIGCHI conference on human factors in computing systems*, ACM, New York, pp. 423-432.

Dittert, N, Katterfeldt, E-S & Schelhowe, H 2012, 'Die EduWear-Umgebung – Wearables konstruierend be-greifen', *i-com*, vol. 11, no. 2, pp. 37-43.

Fab Charter 2012, viewed 11 January 2013, <http://fab.cba.mit.edu/about/charter/>.

Frauenfelder, M 2010, *Made by hand: Searching for meaning in a throwaway world*, Portfolio, New York.

Katterfeldt, E-S & Schelhowe, H 2008, 'A modelling tool to support children making their ideas work', *Proceedings of the 7th international conference on interaction design and children*, ACM, Chicago, pp. 218-225.

Kuznetsov, S & Paulos, E 2010, 'Rise of the expert amateur: DIY projects, communities, and cultures', *Proceedings of the 6th Nordic conference on human-computer interaction: extending boundaries*. NordiCHI '10. ACM, New York, pp. 295-304.

Millner, A & Baafi, E 2011, 'Modkit: blending and extending approachable platforms for creating computer programs and interactive objects', *Proceedings of the 10th international conference on interaction design and children*, ACM, New York, pp. 250-253.

Mota, C 2011 'The rise of personal fabrication', *Proceedings of the 8th ACM conference on creativity and cognition*, ACM, New York, pp. 279-288.

Muutanen, U-M 2005, *Crafter Manifesto*, no. 4, p. 7.

Paulos, E 2012, 'You amateur!', *interactions*, vol. 19, no. 1, pp. 52-57.

Resnick, M et al. 2005, *Design Principles for Tools to Support Creative Thinking*, Carnegie Mellon University, Pittsburgh, viewed 19 March 2010, <http://repository.cmu.edu/isr/816>.

Resnick, M & Silverman, B 2005, 'Some reflections on designing construction kits for kids', *Proceeding of the 2005 conference on interaction design and children*, ACM, New York, pp. 117-122.

De Roeck, D, Slegers, K, Criel, J, Godon, M, Claeys, L & Kilpi, K 2012, 'I would DiYSE for it!: a manifesto for do-it-yourself internet-of-things creation', *Proceedings of the 7th Nordic conference on human-computer interaction: Making sense through design*. NordiCHI '12. ACM, New York, pp. 170-179.

Torrey, C & McDonald, DW 2007, 'How-To Web Pages', *Computer*, vol. 40, no. 8, pp. 96-97.

THOUGHTS FROM THE ROAD OF A FABLAB NOMAD

JENS DYVIK

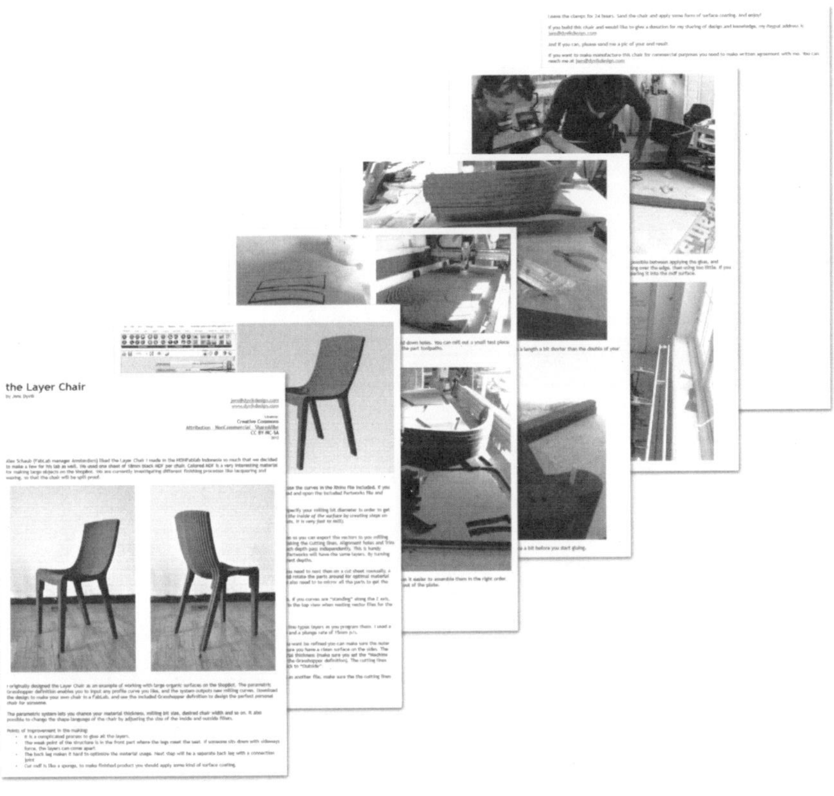

Figure 1: Online documentation (Source: Design by Jens Dyvik).

1. FabLabs and Handicraft

I am currently on a FabLab world tour. My goal is to research personal manufacturing and Open Source design. One of my main fascinations with the FabLab world is how handicraft and production by machines come together. Based on my experiences, in this article I describe the potentials FabLabs offer for design and the challenges FabLabs have to face in the future.

1.1 The Mind and the Hand

Since our industrial revolution, the design, manufacturing, and use of products have become separated. As a consequence, a product's theoretical development, its production and its use are usually carried out by people who have nothing to do with each other. However, I believe that the more a person knows about the story behind a product, i.e., how it was designed and how it was made, the more will this person cherish the creation.

By enabling people to create products with digital machines, FabLabs help to create stronger connections between people and things. There are a lot of positive consequences, both practical and philosophical ones, from people making the products and tools they need in their own lives. When people make tools for their own needs, the line between creation and implementation is very small. Moreover, creating the physical 'ingredients' of your life, or of the life of someone you care about, can be very rewarding.

1.2 Far More than Pressing a Button

Making a product with a digital machine like a CNC mill or a laser cutter involves far more than just pressing a button. First of all, your idea needs to be converted into a digital drawing. Then, material needs to be chosen and prepared. Once the parts of a design have been produced by the machine, they need assembly and finishing. The combination of abstract design by software- and computer-controlled machines with physical preparation, operation of CAM tools, assembly and finishing by hand, can be very valuable.

Since the design is made with the help of computer software, it becomes possible to create precise designs without years of handicraft training. But you can only make what you can draw digitally in a FabLab. Open Design and the sharing of design files can be of great help with this. Digital design means that you can load the creation of another FabLab user into your design software and personalize or add to this design without having to start from scratch. Like this, new users of digital fabrication get a flying start, and experienced users can collaborate more easily and reach new design combinations.

But still the hands-on modification of your design is important. Maybe once you have your laser-cut design in your hands, you realize that you actually wanted a corner to be rounded and you wish to sand it off. Like this, users can grind, saw and glue their way to a finalized design. Likewise, the choice of finishing like lacquer or paint and the combination of materials are examples of how craftsmanship meets computer work. FabLab machines are also used to create tools for handicraft instead of a finished product, such as weaving looms or tools for painting.

1.3 The FabLab Network as a Collective Creative Force

Another advantage of making products using digital machines is that the manufacturing process can be repeated with very little effort. This means that once you have converted an idea into a computer drawing, you can easily make several variations in proportions, size and so on. These digital designs can easily be shared by transferring the bits and bytes behind them. When these continued iterations of design flow between the FabLabs around the globe, the FabLab network works as one collective creative force. The process of continued iterations of a design can also involve a lot of play. This playfulness and search for new possibilities is one of the positive consequences you get when creators operate the machines that make their own creations.

Converting an idea into a digital drawing is not necessarily easy though, and this is the skill I most commonly get asked to share when I visit a FabLab. Using a FabLab requires you to master the craft of operating a computer to a certain degree. For some users from younger generations, bits and bytes are flowing in their blood from childhood and new programs and possibilities are learnt in the blink of an eye. For other users though, getting to a finished digital drawing of a design and sending this to a machine can be a fight of terror and frustration. And when it really goes wrong, the benefits of digital manufacturing and knowledge sharing are outweighed by the struggle to operate the computer. This is why easy-to-learn yet powerful design software is crucial to the FabLab world. Luckily, there are improved tools being released every year, both by the community and by external companies. However, it is more important to enable users to understand how to use the programs. Workshops and courses are great for this, but online resources and the mapping of these are also very valuable.

The accessibility of software is a very important ingredient for creating a thriving FabLab community. Expensive software can be a major blocking factor, especially in lower income countries. This is why Open Source software is crucial to the FabLab world. The downside is that Open Source software can be more difficult to learn and that it may not include design tools that are adequate to commercial software. I personally tend to use a non-Open Source design program for this reason. A consequence of this can be that only users who have purchased or hacked this program can build upon or benefit from my designs. To avoid this, I always publish my designs in a format that you can import into an Open Source program next to the original design file. I am hoping to switch to Open Source software only once the tools I use for my designing become available. The software used for controlling the FabLab machines is becoming easier and easier to use, and we are moving towards a future where you only need to operate a single software to both design and make your parts in the FabLab.

2. Tools for Local Manufacturing and Tools for Global Collaboration

However, we do not only need software for design and manufacturing – also for the collaboration between people. In the following section I describe various examples of how people share knowledge and experience supported by digital media (applications).

2.1 Variations in Software

The FabLab's mantra is to collaborate globally and produce locally. We have seen great advances in the development of tools for design and manufacturing, but the FabLab community also needs better tools for collaboration and knowledge sharing between the labs. FabLab is all about people collaborating and sharing knowledge, and I think in the coming years we will see some great new infrastructures for doing this. The polycom video conferencing system that allows the FabLabs to talk to each other may be the most successful collaboration tool so far. Yet this tool is still far from being used at its full potential.

Most FabLabs do very well in sharing knowledge and empowering users locally. Making the knowledge created in one lab available to other labs in the network often does not work very well though. It is all about the documentation of what and how something was made and publishing this. Most FabLabs do a fairly bad job at this so far. There are some highlights though, like Waag Society's FabLab in Amsterdam who have a very large library of documented work by their users on their website.

When users create something in a FabLab and share the knowledge of how they did it, you get a beautiful convergence of creation and assisting others in creation. These libraries of fab moments become a valuable addition to more formal teaching sources like books, lectures and courses. Traditionally, you would have one professional spending years perfecting the craft of making something and then publishing a book about it. Now we also have professionals and amateurs collaborating and documenting their work online as they go. I believe that this enables a more free-flowing way of applying the knowledge of others in your work. And I believe this democratized knowledge makes it possible for more people to create better and more valuable things.

2.2 Applying Shared Knowledge

Once knowledge and creation have been shared, the next challenge is to apply this knowledge somewhere else. Surprisingly seldom we see that users of one FabLab replicate or innovate a creation coming from another lab. An example is a very well designed 'foosball table' with automatic score count and replay camera from FabLab Amsterdam.

The files for making this soccer game are available online as well as instructions for making it. The production involves a lot of core FabLab techniques and it is an ideal learning project to become an advanced FabLab user. As of right now, no one

Figure 2: Illustration and photography of the 'foosball table' (Source: Waag Society).

has replicated this Open Design, and reasons for this lack of imitation may vary. Maybe the design is not exposed enough, and the people who could benefit from this knowledge do not know it exists. Or maybe the materials and time needed make it too expensive. I think there is a lot more indirect knowledge absorption going on though. Maybe someone saw the documentation and applied some of the design or manufacturing techniques in their own project. Mapping this kind of direct and indirect application of shared knowledge could be of great value. A knowledgeable transfer-mapping tool might mark an important part of future collaboration infrastructures in the FabLab community.

3. FabLabs as Social Hubs

Beside the flow of knowledge and collaboration between the labs, the most powerful interaction between people happens within a lab. In the last section I depict several other aspects that are important for FabLabs being a social hub.

3.1 The Power of Face-to-Face Interaction

FabLabs are social hubs that connect people and ideas. When a FabLab functions well, the accessibility, openness and freedom to try out ideas brings out the best in people. This atmosphere becomes contagious and people willingly share ideas, techniques and knowledge with each other. It is as though a switch gets flipped in a visitor's head from a protecting to a sharing spirit. A user who comes to the lab with one specific goal might end up going home with a lot of unexpected inspiration and new collaboration partners. The space for failure is also very important. Unlike many traditional knowledge institutes, failure is welcome in a FabLab as long as you make sure you and others learn from it. New users to a complex machine are bound to make many mistakes, but the mistakes lead to deeper insights into the workings and possibilities of a technology. If you get a user of a FabLab so far that he or she feels welcome to experiment and try out new possibilities, you have started to unlock a very powerful method for innovation.

3.2 More than Making

This facility as a social service goes deeper than helping people meet and inspire each other. Some successful FabLabs become safe havens for individuals that feel lost in society at a given moment. The warm-hearted inclusiveness and freedom to express yourself in a lab can play a supporting and healing role in an individual's life. This is a very positive result of an open workspace full of passionate people. MIT FabLab Lyngen is one specific lab with many examples of this function, but there are success stories to be found in many other labs as well. These social and indirect functions of FabLabs are rarely mentioned anywhere. Still I find this one the most fascinating and important aspects of FabLabs. When it comes to justifying the cost of running a FabLab and the quest for a sustainable business model, I think these more hidden benefits need to be taken into consideration.

Moreover, the potential of FabLabs to improve the daily life of citizens in developing countries is quite obvious. By enabling people to create the tools they need where they live, FabLabs have already improved the lives of many poor people. An example of this kind of effort is the '$50 prosthetic leg program' (see also the article by Alex Schaub in this book) of HONF FabLab Yogyakarta, Indonesia and Waag Society's FabLab Amsterdam, Netherlands. As a part of my FabLab travels, I have contributed with some design ideas and workflows to its goal to locally manufacture an affordable prosthetic leg in Indonesia. In most industrialized countries though, the everyday problems tend to be less physical, as almost all basic needs like shelter, food and mobility are provided for the majority of the population. And if individuals do not lack access to basic needs, they also do not yearn for designed practical solutions but rather for other improvements, such as in the social field. I find that many of the actions and programs in FabLabs have important mental consequences. In a way, the social interaction of creating something physical in the lab can have more consequences than just the physically present product. The physical action and collaboration process of making becomes a vehicle for serving mental needs. The experience of being able to make the things you dream of, to participate in a global movement of collaboration, and to empower your life or that of someone else is really valuable.

This brings me back to my starting point about the connection between people and things. When you make something for yourself, or when somebody makes something for you in a FabLab and you know about the process behind it, this product will mean a lot more to you. If it breaks, you will hopefully want to repair it, and wish to replace this item with a new one later on. This change in attitude towards the value of a physical product may lead to more sustainable lifestyles with less consumption of more cherished items.

I am looking forward to seeing FabLabs developing tools for more sustainable and meaningful living, and to seeing more and more projects and ideas being adapted and improved between the labs. FabLabs are places that utilize modern technology to make the mind and the handwork together in one space again. And this is a place I want to be.

Acknowledgement

I would like to thank the Keep an Eye Foundation (NL) for supporting parts of my research, as well as all the great people who have hosted me at their FabLabs so far.

TECHNOLOGY AND INFRASTRUCTURE

NOTES ON TECHNOLOGY AND INFRASTRUCTURE

BRE PETTIS

Figure 1: 3D printer (Source: Photography by FabLab Aachen).

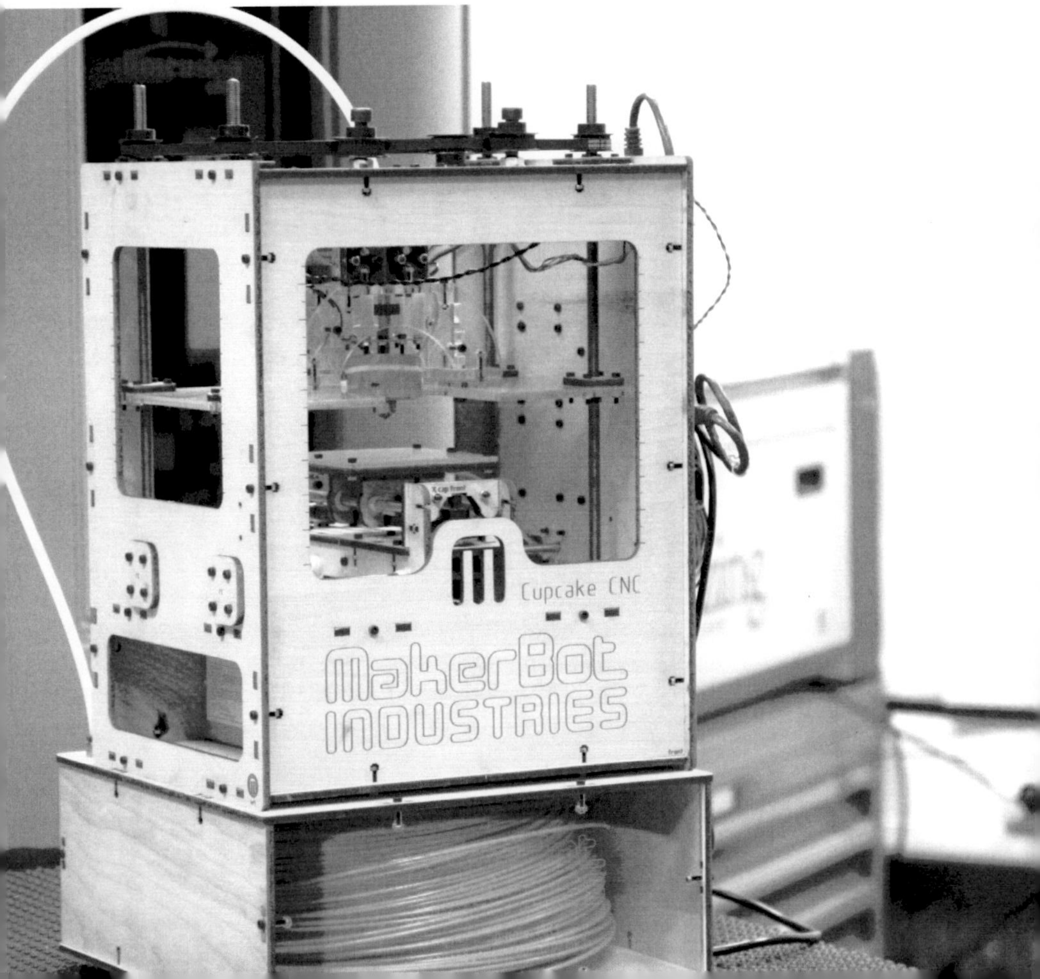

My name is Bre Pettis from Makerbot®. Founded in 2009, Brooklyn-based MakerBot (www.makerbot.com) has grown to be a global leader in desktop 3D printing. MakerBot had 16% market share of all 3D printers (industrial and personal) made from 2009 to the end of 2011 (Wohlers Associates 2012). There are over 13.000 MakerBot Desktop 3D Printers in use by engineers, designers, researchers, and people who just like to make things. MakerBot was named one of the top 20 startups in New York City and has been featured in The New York Times, The Wall Street Journal, the Economist, Wired, The Colbert Report, Fast Company, Engadget, Make: Magazine, Rolling Stone, Time.com, IEEE Spectrum, CNN, Financial Times, NPR, and many others.

MakerBot 3D Printers are available for sale through www.makerbot.com/store, its worldwide network of distributors, as well as through the company's brand new flagship retail location. Located in the NoHo district of Manhattan, the MakerBot Store offers people the chance to learn about desktop 3D printing, get demonstrations of MakerBot's latest desktop printer, the MakerBot Replicator 2 Desktop 3D Printer, and buy all kinds of new consumer products made on MakerBot Replicator 2 Desktop 3D Printers.

The MakerBot Replicator 2 Desktop 3D Printer is the company's fourth-generation machine, and sets new standards for resolution, build volume, speed, ease of use, and affordability. Hence, the MakerBot Replicator 2 is perfect for FabLabs. Some analysts following the additive manufacturing market believe that 3D printers are having a greater impact on manufacturing than any other technology. 3D printing makes product design and innovation easier and faster – e.g., at home and in FabLabs. This printing type improves product quality, and reduces costs as well as time to market. Moreover, 3D printing enables individual makers and small businesses to manufacture products in small quantities to start new businesses and create jobs. As a result, the additive manufacturing and 3D printer markets are constantly growing. Moreover, 3D printing enables companies and entrepreneurs to make customized products for themselves or for customers in limited run productions. In this way, MakerBot is doing its part to create more jobs for people all over the world.

In the future, we envision a 3D printer in every company and in every home – and we are positioning ourselves to be the global leader in desktop 3D printing. At this point, we would compare the MakerBot Replicator 2 Desktop 3D Printer to the Apple II. In Star Trek, Captains Kirk and Jean-Luc Picard would tell the Enterprise Replicator to make everything from tea to spare parts. We are not sure about the tea, but we are confident that in ten until 20 years, MakerBot 3D printers will be able to make spare parts virtually on demand.

Furthermore, at MakerBot's Thingiverse website (www.thingiverse.com) all 3D designers can access and contribute to more than 28.000 projects, models, and things that can be downloaded and made for free. Thingiverse is a place for engineers, designers, researchers, and anyone who likes to make things to share their digital designs with the world. MakerBot believes that just as computing shifted away from the mainframe into the personal computer that you use today, digital fabrication will share the same path. In fact, it is already happening: laser cutters, CNC machines, 3D printers, and even automated paper cutters are all getting cheaper by the day. These machines are useful for a huge variety of things, but you need to supply them with a digital design in order to get anything useful out of them. We are hoping that together we can create a community of people who create and share designs freely, so that all can benefit from them.

Reference

Wohlers, T 2012, *Wohlers Report 2012 - Additive Manufacturing and 3D Printing State of the Industry*, Wohlers Associates, Fort Collins, Colorado.

MACHINES FOR PERSONAL FABRICATION

RENÉ BOHNE

1. INTRODUCTION

Digital fabrication is a process that can make a solid object from a digital model. Tools for digital fabrication are computer-controlled machines like CNC milling machines, laser cutters, plotters, and 3D printers. And FabLabs in turn are open workshops equipped with these digital fabrication tools. In theory, everybody can create almost anything in a FabLab (Gershenfeld 2005). In reality, people need help from trained technicians and designers, because the devices are not easy to use, and the software tools are mostly created for professionals. In the future, the digital fabrication tools might move into every household – this scenario is called personal fabrication. This article explains several tools and processes for digital fabrication as so used in FabLabs.

2. DIGITAL FABRICATION

This section explains basic digital fabrication processes that have been used in numerous industries for many decades. This section also focuses on processes that are used in FabLabs. Two kinds of digital fabrication processes exist: additive fabrication and subtractive fabrication. CNC routers and laser cutters are devices that remove material from a big sheet or block of raw material – they subtract material. In contrast, 3D printers create objects by adding material in an additive fabrication process.

2.1 Why Digital?

Digital fabrication uses digital models that are stored in files. It is easy to share and copy digital files, because no information is modified or lost in these processes and errors are easy to detect. Sending files over the Internet is very cheap compared to the shipping costs for prints or prototypes. Algorithms and computer programs can create and modify digital data at high speed and with high precision. It is possible to print the same object many times, using the same digital model. The model does not wear out and does not lose details during the fabrication process. Consequently, digital files are the foundation for cheap, fast, precise, and easily reproducible fabrication.

2.2 Subtractive Fabrication

Figure 1: LPKF Protomat S62 CNC milling machine for PCBs. This machine can change ten tools automatically (Source: Photography by FabLab Aachen).

Subtractive fabrication processes remove material from a sheet or block of raw material. Brown (1991) gives a good introduction into the many manufacturing processes. For this article I concentrate on the processes that are available in most FabLabs today: machining tools and laser cutters.

Both CNC milling machines and CNC routers use rotating tool heads. The tool heads hold machining tools that remove material from the work piece. The tools have to be replaced after some time since the milling process damages them. Most CNC milling machines can be used for metal, while most routers just cut or engrave sheets of wood and plastics. Small and precise machines can create Printed Circuit Boards (PCBs) that are used in electronics. CNC milling machines and CNC routers are available in all sizes.

Data for the CNC machine has to be created with Computer Aided Design (CAD) software. With Computer Aided Manufacturing (CAM) software, the designs are prepared for the CNC machine. The user has to select the tools for a given path or layer, set the depth, speed, and many other parameters that must fit the material. The CAM software creates a sequence of machine dependent commands in the G-Programming Language (G-Code) and sends it to the CNC machine. The G-Code controls the movement of the device.

Another form of subtractive fabrication uses laser cutters. A laser cutter is a device that uses a laser for cutting or engraving materials. Different lasers are used for different materials. The laser beam hits the surface of the work piece and depending on the power and material properties, the beam can remove material from the surface. Lasers are used in many manufacturing processes today. Ion (2005) gives a good overview of these processes. Most FabLabs use carbon dioxide (CO_2) laser cutters with up to 100W that fit on a desktop. While 100W is enough to cut sheets of materials like paper, wood, or acrylic, industrial machines with many thousands of watts are able to cut even strong alloys like steel. A typical laser cutter in a FabLab has a working area of 600mm · 300mm. Most expensive models come with a printer driver for Microsoft Windows. Operating a laser cutter with Linux or Mac OS X was very difficult in the past, but today it is possible to use almost any operating system, because the VisiCut[1] software (a user-friendly tool for laser-cutting developed at our local university) is available for many laser cutters.

In general, two parameters control the cut: speed and power. High speed and low power result in low energy per time unit and area. Low energy is good for scratching surfaces or for cutting very thin materials without burning or melting them. Slower speed and higher power result in more energy per time unit and area. The laser beam stays longer on any given point of the surface and has high energy, so it can cut thicker materials. But there is a risk that the material burns or melts or that at least the cut will become ugly because of residues. Most laser cutters support two operating modes: vector mode and raster mode. In vector mode, the laser head follows a given vector path that is stored in a vector image file. In raster mode, the laser head moves quickly from left to right, one row after the other, and fires if a point is present at the given x/y coordinate in the corresponding raster bitmap file. Usually, raster mode is used for engraving pictures or text, while vector mode is used for cutting shapes; but it is also possible to use vector mode with low energy to engrave a vector path. Laser cutters are 2D machines. The laser hits the surface from above. Some machines have a computer-controlled z-axis and can move the material during the process. This allows 2.5D engraving at a high resolution.

2.3 Additive Fabrication

Among the first applications for additive fabrication were prototypes for the automotive industry. Prototyping is the most important application field for additive fabrication today, but it is also used in small-quantity production systems. For higher quantities, many additive processes of today are too slow, but they can

1 | For more information see the website, last viewed 20 December 2012 <http://hci.rwth-aachen.de/VisiCut>.

be used for building molds for faster, established processes for mass production. A detailed overview of all additive manufacturing technologies can be found in (Gibson 2009). The most important additive fabrication technologies are: Stereo Lithography (SLA), Selective Laser Sintering (SLS), 3D Printing (3DP), Fused Deposition Modeling (FDM), and Laminated Object Manufacturing (LOM).

The oldest process in additive fabrication is called Stereo Lithography (SLA). It was invented in the 1980s by Charles W. Hull (1986). The first machines used ultraviolet (UV) light and liquid photopolymer. A focused beam of UV light draws one layer of the 3D model onto the surface of the liquid. The liquid cures after it got exposed to the UV light and joins with the previous layer of the object. This process is repeated layer by layer, until the whole object is created. The finished objects are solid and they are strong enough to be machined. It is possible to use them as master patterns for processes like injection molding, blow molding, thermoforming, and even for some metal casting processes.

The Selective Laser Sintering (SLS) process is similar to SLA. Instead of a liquid polymer, SLS uses plastics, metal, glass, or ceramic powders. A CO_2 laser is used instead of the UV light. The laser fuses the powder together. The final objects that SLS produces are fully functional. 3D Printing (3DP) is similar to SLS, but instead of a laser, liquid binding material binds the powder together. Some machines even add ink to the liquid, which allows printing multi-colored 3D objects. The printing heads are very similar to those used in inkjet printers. Fused Deposition Modeling (FDM) machines are computer-controlled hot glue guns. They heat up and extrude plastics and put it layer by layer on the platform. The layers glue together after cooling and the object stays solid. Because the layer resolution of most machines is worse than SLA or SLS, the surface quality of the objects is not perfect. Another disadvantage is that support material has to be added if the model contains overhangs. The support material is a support for the model material and it has to be removed after the process.

Figure 2: Gold ring, created using SLS (Source: Photography by FabLab Aachen).

Laminated Object Manufacturing (LOM) uses a CO_2 laser or knives to create an object from layers of paper or plastics with a special coating on the back. A heated roller presses one layer of coated paper to the previous layer. The CO_2 laser then cuts the top layer into shape. Unused areas of the top layer can be cut into small rectangles. They function as support material for the next layer. When the construction is completed, all support materials must be removed by shaking the object.

3. Digital Fabrication in FabLabs

In the beginning, FabLabs were equipped with the same machines that are used in industrial environments. Today, FabLabs also build their own machines and they also buy low-cost devices used by hobbyists. One goal of FabLabs is to make the digital fabrication tools available to everyone. A brief overview of the most common machines is given on the FabWiki-Equipment page[2], while I concentrate on the core equipment for fabrication:

- Computer controlled subtractive machining: Laser cutter (e.g. Epilog Laser), CNC large-scale wood router of soft materials mill (e.g. ShopBot), vinyl/sign cutter (e.g. Roland CAMM), precision desktop milling machine (e.g. Roland Modela).
- Computer controlled additive machining: UP! (Allegedly to be replaced by the Ultimaker when it can be rolled out in volume).

The second reference is the FabLab Inventory List. Originally, this was a simple text file located at fab.cba.mit.edu/about/fab/inv.html; the page now redirects to a google document[3] and a lot of the inventory items have been changed. While the professional laser cutter and the big CNC router are still in the new document, the expensive FDM printers were removed and replaced by mid-price models. This shows that the FabLab inventory is not static but changes with the needs of the community. Many new FabLabs do not buy expensive 3D printers, but they start with a mid-priced or a cheap DIY 3D printer for personal fabrication (see next section).

The FabLab conformity rating (2012) distinguishes three levels for the tools and processes, depending on the equipment of the FabLab. The best rating (A) is given if a FabLab has "all core tools and processes and possibly more" (FabLab conformity rating 2012) according to the third column ("common set of tools and processes" (FabLab conformity rating 2012)) of the FabLab conformity rating. The (B) rating is applied if a FabLab has "very close to but missing at least one core machine or process" (FabLab conformity rating 2012). And the FabLab conformity rating gives the worst rating (C) to a FabLab if it is "difficult to do most fab projects or follow fab tutorials" (FabLab conformity rating 2012).

3.1 Machines for Additive Fabrication in FabLabs

Almost all A-rated FabLabs have a high-priced FDM 3D printer from Stratasys (other brands: Dimension, HP). A few FabLabs (e.g., Protospace Utrecht and FabLab Barcelona) use a Zcorp Z-Printer. In addition, almost all FabLabs have at least one low-cost DIY 3D printer.

2 | For more information see the website, last viewed 20 December 2012 <http://wiki.fablab.is/wiki/Portal:Equipment>.

3 | For more information see the website, last viewed 20 December 2012 <https://docs.google.com/spreadsheet/pub?key=0AtIlZyLn99e6dGRleUJTY043a3FucUhFUVVBYTdxS3c&single=true&gid=0&output=html>.

3.2 Machines for Subtractive Fabrication in FabLabs

Most FabLabs have an Epilog[4] laser cutter. The most famous model is the Mini24, but some FabLabs use Epilog Zing laser cutters. Some other laser cutters that are used include Trotec[5] Speedy 300 (Groningen) and LaserPro[6] Spirit GE (Amsterdam, Barcelona, Luzern). Almost all FabLabs have a small milling machine. The most famous model is the Roland Modela MDX-20[7]. Most FabLabs have a Shopbot[8], too. Other big dimension CNC milling machines or routers that are used in some FabLabs are machines by BZT[9] (Amsterdam, Luzern), Seron 1325 (Protospace), and Precix 3000[10] (Barcelona). Almost all FabLabs have a vinyl cutter from the company Roland. The most famous cutter is the Roland CAMM 1 GX-24.

4. Personal Fabrication

In contrast to industrial digital fabrication devices that can cost at least 10.000€, the digital fabrication tools that people will use at home need to be much cheaper, so that everybody can afford a 3D printer or a laser cutter. They also need better software that is easy to use. This section introduces some machines that are available today or in the near future. They can be used at home or in FabLabs in addition to the high-priced machines that we find there today.

4.1 3D Printers for Personal Fabrication

Adrian Bowyer started a revolution with the RepRap[11] (Replicating Rapid Prototyper) project. The design files of this 3D printer and the source code for its software were released under Open Source licenses. The original idea of a self-replicating machine created a community of Open Source enthusiasts that forked the source files and invented their own 3D printers. Malone (2007) proposes that an Open Source, low-cost, personal 3D printer kit can accelerate the spread of the technology.

4 | For more information see the website, last viewed 20 December 2012 <http://www.epiloglaser.com>.

5 | For more information see the website, last viewed 20 December 2012 <http://www.troteclaser.com/>.

6 | For more information see the website, last viewed 20 December 2012 <http://www.laserprona.com/>.

7 | For more information see the website, last viewed 20 December 2012 <http://www.rolanddga.com/>.

8 | For more information see the website, last viewed 20 December 2012 <http://www. shop-bottools.com/>.

9 | For more information see the website, last viewed 20 December 2012 <http://www.bzt.cnc.de>.

10 | For more information see the website, last viewed 20 December 2012 <http://www.precix.com/>.

11 | For more information see the website, last viewed 20 December 2012 <http://www.reprap.org>.

The system was called Fab@Home Model 1 and the project did not just contain the hardware but also a network that supports the development of this system. Back in 2007, the RepRap project did not offer hardware kits that contained all parts needed for a 3D printer. A company called Makerbot Industries addressed this problem (see also the article by Bre Pettis in this book). The Cupcake CNC 3D printer is based on early work from the RepRap project and the company sold this product since 2009 as a kit with all parts that are needed in order to build a 3D printer at home. Their second product was the Thing-O-Matic and in 2012 they released their latest product: the Replicator 2. The new model is more expensive, but it has a higher build quality and it comes fully assembled. Another important contribution of Makerbot Industries is their Internet platform called Thingiverse[12], where users can upload and share 3D models and other Open Source hardware projects. Since 2011, also the Dutch company Ultimaking Ltd. sells 3D printers. The first model is called Ultimaker and it was developed in the Protospace Utrecht FabLab. Like many other DIY 3D printers that appeared on the Internet in the past years, the Ultimaker is based on the RepRap project.

One low-cost FDM 3D printer is very famous in FabLabs although it is not an Open Source project: the UP! by a company called PP3DP[13]. It is a fully assembled machine and not a kit. All low-cost printers currently use FDM technology, as it is cheap and easy to implement. But the Open Source hardware community works on machines that use SLA technology. The B9Creator[14] is a DIY 3D printer using SLA technology. The user has to assemble the machine. A commercial low-cost SLA printer is the Form 1, created by a company called formlabs[15].

4.2 Laser Cutters for Personal Fabrication

The Lasersaur[16] is an Open Source hardware laser cutter. It has overall dimensions of 1700 mm · 1080 mm · 350 mm and a work area of 1220 mm · 610 mm. The problem with this project is the fact that lasers are dangerous and users have to assemble the Lasersaur on their own. Some parts are hard to get and in some countries, people need special training and certificates to buy some dangerous components.

Moreover, there are many low-cost laser cutters on the market. They are fully assembled and have a closed housing, just like the professional laser cutters that are used in most FabLabs. One of the most famous systems comes from the UK,

12| For more information see the website, last viewed 20 December 2012 <http://www.thingiverse.com>.

13| For more information see the website, last viewed 20 December 2012 <http://pp3dp.com/>.

14| For more information see the website, last viewed 20 December 2012 <http://www.b9creator.com>.

15| For more information see the website, last viewed 20 December 2012 <http://formlabs.com/>.

16| For more information see the website, last viewed 20 December 2012 < http://labs.nortd.com/lasersaur/>.

Figure 3: HPC-3020 laser cutter. LAOS electronics can be added which make it compatible with VisiCut (Source: Photography by Peter Brier).

produced by a company called HPC[17]. It is famous in FabLabs because it is very cheap and some models, like the HPC-3020, are compatible with the Laser Open Source (LAOS)[18] project. LAOS modifies cheap laser cutters like the HPC-3020. New electronics add more features to the low-cost devices. The LAOS board is Open Source hardware and its firmware is Open Source software. It is compatible to the VisiCut software and therefore, LAOS-enabled laser cutters can be used with any platform.

4.3 CNC Mills for Personal Fabrication

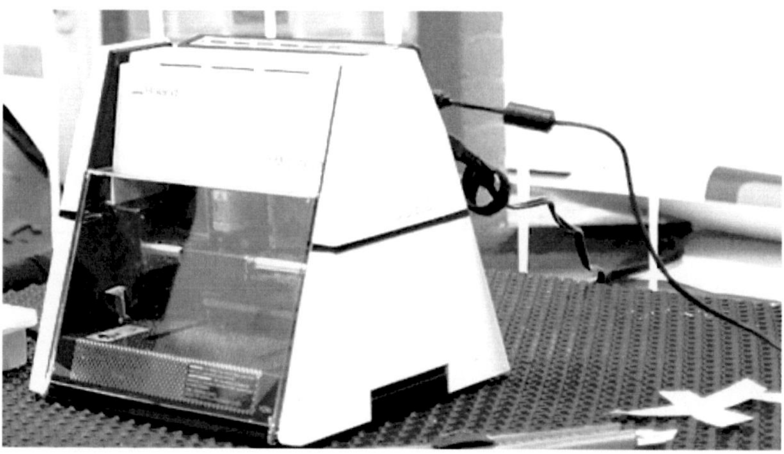

Figure 4: Roland iModela CNC milling machine (Source: Photography by FabLab Aachen).

17 | For more information see the website, last viewed 20 December 2012 <http://hpclaser.co.uk/>.

18 | For more information see the website, last viewed 20 December 2012 <http://www.laoslascr.org/>.

The Mantis Mill[19] is a low-cost DIY CNC machine. The current design has 13 parts, all of which can be made with a handsaw and a drill press. The goal of this project was to build a 3-axis CNC milling machine for less than $100. It is part of the MIT Machines that Make (MtM)[20] project. MtM has many low-cost machines that can be built in a FabLab. A low-cost commercial CNC milling machine is the Roland iModela. It is a very small (86 mm · 55 mm · 26 mm) device that can be used for soft materials. First users created small PCBs on this machine.

5. Conclusion

In this paper, I gave a brief overview of additive and subtractive digital fabrication technologies that are used in FabLabs today. Since there is not a mandatory list with machines that every FabLab needs to have, I decided to name the most famous devices. It also seems that FabLabs increasingly build their own machines so that they do not rely on the commercial products that were developed for industrial applications anymore. The next generations of these low-cost DIY machines might make personal fabrication at home a reality. Even today, commercial 3D printers like the Replicator2 and the UP! can be used at home and people can use them with less technical skills than even five years ago.

References

Brown, J 1991, *Modern Manufacturing Processes*, Industrial Press Inc, New York.

FabLab conformity rating 2012, last viewed 20 December 2012, <http://wiki.fablab.is/wiki/Fab_Lab_conformity_rating>.

Gershenfeld, N 2005, *Fab. The coming revolution on your desktop – from personal computers to personal fabrication*, Basic Books, New York.

Gibson, I & Rosen DW & Stuckler B 2009, *Additive Manufacturing Technologies: Rapid Prototyping to Direct Digital Manufacturing*, Springer, New York et al.

Hull, CW 1986, *Apparatus for Production of Three-Dimensional Objects by Stereolithography*, US Patent 4575330.

Ion, JC 2005, *Laser Processing of Engineering Materials: Principles, Procedure and Industrial Application*, Butterworth-Heinemann Ltd., Amsterdam et al.

Malone, E & Lipson, H 2007, 'Fab@Home: The Personal Desktop Fabricator Kit', *Rapid Prototyping Journal*, vol. 13, no. 4, pp. 245-255.

19 | For more information see the website, last viewed 20 December 2012 <http://makeyourbot.org/mantis9-1>.

20 | For more information see the website, last viewed 20 December 2012 <http://mtm.cba.mit.edu/>.

DIGITAL FABRICATION IN EDUCATIONAL CONTEXTS

IDEAS FOR A CONSTRUCTIONIST WORKSHOP SETTING

NADINE DITTERT, DENNIS KRANNICH

1. Introduction

In his book Fab – The Coming Revolution on Your Desktop. From Personal Computers to Personal Fabrication, Neil Gershenfeld (2005) describes Seymour Papert's ideas as bringing together the worlds of play and work by blurring the distinction between toys and tools for invention. FabLabs follow this idea and hold the potential to take it to another level, making design and engineering more appealing and more transparent at the same time. Hence, digital fabrication enables us to explore how to represent a functional description of a system in physical form and likewise to which extent a functional description of a physical system can be abstracted.

The construction process may still start with hand-drawings (they are still very important for brainstorming and rapid-visualization), but the main work is done on the computer. We use CAD (Computer Aided Design) or vector programs for exact constructions and can easily print them out two- or three-dimensionally, using different materials ranging from paper over wood and acrylic glass to ABS- or PLA-plastic. Additionally, computer simulations help us to analyze material strength and weight. The computer also helps us to calculate the optimal balance between stability and weight.

Since computers and software are becoming more and more affordable, we can now implement our ideas, even in our own homes. We can construct things that we have dreamed of and that solve our particular problems. We do not have to accept what others have built for us: Now we can build what we need. We do not depend on professional companies anymore: We can now produce single and unique artifacts instead of small series or mass productions. We can, for instance, print out connectors that connect the various types of Thomas Railway Systems (Thingiverse 2013), design personalized coasters or invent new goods. In general, digital fabrication encourages a faster and affordable implementation of ideas.

Additionally, FabLabs enable bottom-up processes that allow the community to determine the priorities and needs rather than determining a market opportunity, for instance.

As a consequence of this, the virtual and the real product now are closely linked, enabling us to be more creative and to focus on constructing innovative objects. Now we can step out of the computer and make the virtual objects 'be-greifbar' (graspable and tangible) (Robben & Schelhowe 2012; Krannich et al. 2012). Our assumption is that these close links offer new chances for educational purposes to make design and engineering more transparent. The transformations from hand drawings to 3D models and real products become graspable in a new creative way. People that are willing to learn can experience and understand the entire process of developing products.

In this article, we want to illustrate an example of how to apply digital fabrication technology to an educational context. First, we describe a project that was implemented by five undergraduates at the University of Bremen. Then we outline a pedagogical concept for constructionist workshops that has been applied for more than six years with children and young people. Based on these experiences, we will present a first draft of how to combine digital fabrication technologies with 'traditional' constructionist workshops.

2. The 'Fab-tast-O-matic'

In the one-year Bachelor project FabLabs, 15 students were asked to create 'incredible machines' known from the computer game The Incredible Machine by Sierra Entertainment. The challenge was to combine digital fabrication technologies known from FabLabs with digital media.

The project was split into five phases: (1) imagination and rapid-prototyping, (2) general introduction to digital fabrication technology and research, (3) ideation, (4) main development and (5) final presentation, whereas the students just got a basic input from the project's supervisors (researchers) and had to elaborate/research on their own. Following the constructionist learning paradigm by Seymour Papert (1980), the students worked in small teams to create their artifacts.

By using FabLab technologies, we wanted the students to encounter digital fabrication and its underlying concepts. Similar to the interdisciplinary approach of the study program Digital Media, this project required different skills like CAD construction, understanding materials, material strength and physical constraints, programming, designing and tinkering. Digital fabrication technologies provided the setting to empower these skills and allowed for connecting disciplines that normally do not belong together, but have much in common, like production design and informatics.

One 'incredible machine' that derived from this Bachelor project is the 'Fab-tast-O-matic' (see Figure 1) created by five female students. The idea evolved not only from the aforementioned computer game, but also mainly from the remarkable cartoons of absurd and complicated machines that perform very simple tasks depicted by Rube Goldberg (*1883 †1970). Besides this, the group was also inspired by the commercials of Coca Cola from 2007, in which imaginative creatures produce

Figure 1: 'Fab-tast-O-matic' with four different levels
(Source: Photography by Justus Holzberger).

a Coca Cola bottle in an unobstructed, sequential process, and by the music video This Too Shall Pass by the band OK Go (2010), in which a chain reaction with different elements like marble tracks, seesaws, and pulleys takes place.

The 'Fab-tast-O-matic' has the size and shape of a vending machine. It consists of independent parts (levels), which represent a magical, playful and original world combined. As another parallel to the vending machine, a trigger will start the actions: instead of inserting a coin, the chain reaction process starts when a gong is struck.

The first level acts as an interface between the machine and the user. After the gong is struck, a sumo wrestler starts to move and activates the first seesaw, which tilts the second seesaw and sets the figures of the Bremen Town Musicians in motion. After this, the chain reaction of the second level is triggered. A wooden ball runs through the marble track and returns to its initial starting position. Subsequently, the LED cube starts its animation and hands over to the third level. Here, water is pumped into a small container. Once the water level exceeds the balance point, the container tilts and triggers the next reaction, the so-called 'step sequencer'. This system gets an input and processes the corresponding output, which acts as a next input. Thus, the sequence is totally random each time the system is started. When a ball hits the ground plate, an audio feedback is returned. The fourth level consists of a laser harp seesaw. Instead of by strings, the sound is triggered by photocells. While the unicorn moves towards the end of the seesaw, a melody is played. At the very end, a sign with the lettering 'The End' pops up and the chain reaction process terminates. The development of this machine started with a simple sketch, similar to those of Rube Goldberg (see Figure 2).

For the implementation, the students extensively used different digital fabrication technologies. For instance, they printed small components for the seesaw mechanism and the sumo wrestler with the 3D printer and cut and engraved different materials (wood and acrylic) to build the marble track, the characters or the machine's enclosure. In order to be able to do so, they had to learn how to use professional 3D construction software (e.g. SolidWorks) to create their physical objects. Since 3D printers have certain limitations in printing, they had to study the production processes and how the machines work in general to be able to print and cut the desired artifact. They also learned basic craftsmanship as well as how to solder the electronic components and to program the Arduino boards that are, for instance, used for the laser harp and the LED cube.

Soldering and programming are also part of constructionist workshops that have been conducted at the University of Bremen since 2006. Before we bring up a setting that combines the digital fabrication technologies with these workshops, we will briefly describe the underlying concept.

3. Constructionist Workshops for Young People

The construction of an artifact offers the opportunity to externalize and reflect on one's own mental models (Papert 1980). Based on this learning paradigm that is known as Constructionism, we have conducted more than 50 workshops where mostly young people have built different artifacts combining tangible

Figure 2: Initial 'Fab-tast-O-matic' sketch (Source: Project Report).

technology like Arduino with craft materials and developed intelligent clothes, scary monsters, magical machines, etc.. These so-called 'TechKreativ' workshops follow a pedagogical concept that has been developed and readjusted over and over again (Dittert et al. 2008; Katterfeldt et al. 2009). The concept follows five different stages, starting with the participants' imagination and ending with a public presentation of their developed artifacts. In the first stage, different methods like brainstorming or fantasy journeys are used to approach the participants' imagination to a given topic like 'Smart Fashion', 'Halloween' or 'Magic'. The aims of this stage are to get visionary ideas of artifacts as well as to open the participants' minds. It is only after this first stage has been completed that the participants get to know the materials that they will work with later during the workshop. They are shortly introduced to the technology hardware and its programming and they are shown the craft materials that are available. In the following third stage, the ideas of what to construct are generated. Hence, the results of the first stage can be combined with the knowledge that was gained during the second stage. Furthermore, completely new ideas might as well emerge. In the fourth and most time-consuming stage, the participants construct and program their artifact in small teams. At the end, the artifacts are publicly presented. A detailed description of the workshop concept can be found in Dittert et al. (2008).

In this constructionist setting, participants externalize their ideas and can reflect upon them while constructing a physical artifact that is enhanced with computing technology. Here, the purpose of the computer is to make the artifact 'come alive', which is done by programming the micro controller with the suitable programming environment. It enables the participants to 'think' like the artifact and reflect on their ideas about the artifacts' behavior.

The digital fabrication technologies allow for further use of the computer during the construction process. The results from the aforementioned Bachelor project were very promising and suggest applying digital fabrication technologies to our workshop concept with children. In the following parts, we want to propose a concept of how to incorporate these technologies and postulate further research questions that we would like to investigate in our future research endeavors.

4. Digital Fabrication Technologies in TechKreativ - Workshops

In order to incorporate digital fabrication technologies into the TechKreativ workshop concept, we suggest adding another step to the second stage (introduction to technology), in which we particularly introduce the different digital fabrication technologies. These machines need to be operated carefully and correctly and their functionality is quite complex compared to normal handheld tools. In order to create artifacts, the students do not only have to learn how to operate these machines but also how to construct the models. Moreover, during the construction phase, the students will elaborate on the machines' functionality in more detail and understand the underlying concepts of computerized numerical control. The usage of digital fabrication technologies allows the students to design very precise constructions of 3D and 2D models, while also offering new ways to make the development process more transparent and 'be-greifbar'.

One example to understand the production process and the related machines could be the creation of the STOOL (Dutch term for chair). It allows for creating a chair by using few 2D objects that can be created with the laser cutter. First, a hand-drawing of the chair is created until the desired object is visualized. In the next step, the students have to think about its construction until they can create a CAD drawing of the different parts. Finally, the chair is cut and its parts are put together to a 3D object to sit on.

In traditional TechKreativ-workshops, all crafts are done without the computer. Students make use of paper, wood, Styrofoam, or textiles and glue, screw or sew their artifacts (that are further equipped with tangible technology and programmed via the computer). Tinkering is an important part of the workshops since it enables creativity and also addresses many young people. Creating parts of the artifact using FabLab technology opens new ways to creativity, learning and construction. The STOOL described above could be the scaffold of an artifact to create in a TechKreativ-workshop. That way, using this technology can enhance the tinkering part of the workshop. The artifact can be more precise, which might be useful for prototyping innovative real world objects, for instance. The 'Fab-tast-O-matic' is a concrete example of what can be created combining TechKreativ with FabLab technology. While TechKreativ encourages prototyping in a tinkering way, the addition of FabLab technologies allows an understanding of real-world production processes. Hence, depending on the target group, the learning objectives, or the overall aim of the workshop, FabLabs can be regarded as an enhancement of TechKreativ workshops. Embedding it into the pedagogical concept would be the next step to its implementation. As described above, another introduction to technology during the second stage of the workshop would be necessary. A detailed approach needs to be elaborated.

In this regard, the following questions arise, calling for further research:

- How can we make design and engineering more transparent?
- How can we use the transformation from a digital model into a real physical artifact to understand the production process and its related learning matters?
- What kind of tools can we use and how do they enhance the learning process?
- What is the relation between 2D and 3D models, and how can we create 3D objects with 2D parts?

5. Conclusion

In this article, we presented our experiences of using digital fabrication technologies in an educational context with university students. We have explained an example of an artifact that can be created and what can be learned during the construction process. Furthermore, we outlined a constructionist workshop setting and how to enhance it by using this new FabLab technology. Research questions have been derived from this and were postulated. The upcoming work is to further elaborate on our approach by conducting and evaluating workshops. The overall aim is to have a pedagogical concept for integrating digital fabrication technologies into TechKreativ workshops for different target groups.

Acknowledgement

We would like to thank the participants of the bachelor project 'fablabs' for their 'incredible machines' as well as all 'TechKreativ' workshop participants, tutors and supporters. We would also like to express our gratitude to Prof. Dr. Heidi Schelhowe for empowering our research.

References

Arzaroli, R, Brehm, I, Crass, M, Hillebrenner, J, Lass, J, Lorz, F, Schwanbeck, D, Bartsch, H, Freundel, E, Happe, H, Kaltenhauser, A, Schade, F, Goszyk, O, Nazarska, R & Martens, V 2012, 'Fablabs-Projektbericht', *Project Report*, University of Bremen, Bremen.

Dittert, N, Dittmann, K, Grüter, T, Kümmel, A, Osterloh, A, Reichel, M et al. 2008, 'Understanding digital media by constructing intelligent artefacts – Design of a learning environment for children', in ED-MEDIA *World Conference on Educational Multimedia, Hypermedia & Telecommunications*, AACE, Chesapeake, VA, USA, pp. 2348-2358.

Katterfeldt, E-S, Dittert, N & Schelhowe, H 2009, 'EduWear: Smart textiles as ways of relating computing technology to everyday life', in *IDC '09 – Proceedings of the 8th International Conference on Interaction Design and Children*, Como, Italy, pp. 9-17.

Krannich, D, Robben, B & Wilske, S 2012, 'Digital fabrication for educational contexts', in *IDC 2012 – Proceedings of International Conference on Interaction Design and Children*, Bremen, Germany, pp. 375-376.

OK Go 2010, *This Too Shall Pass-Rube Goldberg Machine version-Official*, viewed 30 January 2013, <http://www.youtube.com/watch?v=qybUFnY7Y8w>.

Papert, S 1980, *Mindstorms: Children, computers, and powerful ideas*, Basic Books, New York.

Robben, B, Schelhowe, H (eds.) 2012, *Be-greifbare Interaktionen – Der allgegenwärtige Computer: Touchscreens, Wearables, Tangibles und Ubiquitous Computing*, transcript, Bielefeld.

Thingiverse n.d., viewed 31 January 2013, <http://www.thingiverse.com/thing:15757>.

MAKING THE THIRD INDUSTRIAL REVOLUTION

THE STRUGGLE FOR POLYCENTRIC STRUCTURES AND A NEW PEER-PRODUCTION COMMONS IN THE FABLAB COMMUNITY

PETER TROXLER

1. The Third Industrial Revolution

Within a decade, FabLabs have developed from isolated initiatives to a global network of labs that spans all continents. Despite this fast and tremendous growth – or maybe precisely because of it – the global network struggles to define its form and purpose. That network is supposed to provide operational, educational, technical, financial, and logistical assistance beyond what is available within any single lab. Several institutions have emerged and started to provide portions of this assistance: The Fab Academy plays an important role in education. National and international Fab Foundations support operations and logistics of labs. A global user group is developing in the form of an international association. All these institutions are still in their nascent stage, trying to figure out their remit and scope and how to effectively and efficiently work together.

In this chapter, I will place this FabLab ecosystem in the context of a larger development in society, one that has the potential to disrupt or revolutionize the way products are manufactured; it is nothing less than the next industrial revolution. The roots of the FabLab development are technological, and personal digital manufacturing has an important technical side to it. Undoubtedly, developments in manufacturing technology will play an important role in the next industrial revolution. Yet the main disruption that the next industrial revolution will bring is the disruption of hierarchical systems and the emergence of systems of lateral power. Digital manufacturing in a FabLab is personal, and the FabLab network is primarily a social network with lateral connections established between individuals.

It reflects that new paradigm of the third industrial revolution; FabLabs are places of peer-production. Institutions play important roles in providing support in the network where individual connections are not effective or not efficient. They serve specific purposes and provide specific competences. Within the FabLab network, there will be many of these centres of competence. But as of now, such structures have still to fully develop.

In this chapter, I will first outline the concept of the third industrial revolution and explain how FabLabs relate to it. I will then highlight two specific issues in the development of that revolution: First, the difficulties arising from sharing and collaborative development of hardware as Open Source and how seeing FabLabs as a peer-produced commons could help to resolve that issue; second, the institutional challenge that lateral power structures present and how communities and polycentric systems could provide an answer to that challenge. I will then discuss how FabLabs can help to make the third industrial revolution happen and what I would count as a success in such a revolution. To conclude, I will present a roadmap for the future – a set of five central questions around which development will have to evolve, and guidelines how to go about answering these questions.

Many authors have invoked the next or Third Industrial Revolution: Neil Gershenfeld (2005) wrote about 'Fab. The Coming Revolution on Your Desktop', Chris Anderson (2012) claimed that 'In the Next Industrial Revolution, Atoms Are the New Bits' and added that 'Makers [are] The New Industrial Revolution'. Moreover, Jeremy Rifkin (2011) described 'The Third Industrial Revolution – How Lateral Power is Transforming Energy, the Economy, and the World'.

According to Gershenfeld (2005), "possession of the means for industrial production has long been the dividing line between workers and owners. But if those means are easily acquired, and designs freely shared, then hardware is likely to follow the evolution of software. Like its software counterpart, Open Source hardware is starting with simple fabrication functions, while nipping at the heels of complacent companies that don't believe that personal fabrication 'toys' can do the work of their 'real' machines" (Gershenfeld 2005, p. 21).

For Anderson, "the Third Industrial Revolution is best seen as the combination of digital manufacturing and personal manufacturing: the industrialization of the Maker Movement" (Anderson 2012, p. 41). This evidently has two aspects to it. First, digital tools and equipment are becoming widely used by makers both for designing and for manufacturing products, which makes sharing of and collaborating on designs over time and distances easier. Second, as files can be directly sent to machines for production (direct digital manufacturing), makers are able to use pooled manufacturing resources that are larger in scale than what any single maker possibly could afford.

In Rifkin's view, makers and direct digital manufacturing are not the cause but one effect of the Third Industrial Revolution. The revolution itself is actually triggered by changes in communication infrastructure and energy generation as it was the case for the first two industrial revolutions, which were triggered by the invention of the printing press and steam-powered technology in the 19th century, and by electrical means of communication (radio, television) and electricity (mainly from fossil fuels) as main power source in the 20th century. In the Third Industrial Revolution, "the conventional top-down organization of society that characterized much of the economic, social, and political life of the fossil-fuel based industrial revolutions is giving way to distributed and collaborative relationships in the emerging green industrial era. We are in the midst of a profound shift in the very way society is structured, away from hierarchical power and toward lateral power" (Rifkin 2011, p. 36 et sqq.). Furthermore, for Rifkin, the Third Industrial Revolution includes a shift to green buildings, electric cars – and distributed manufacturing: "a new digital manufacturing revolution now opens up the possibility of following suit in the production of durable goods. In the new era, everyone can potentially be their own manufacturer as well as their own power company. Welcome to the world of distributed manufacturing" (Rifkin 2011, p. 117).

1st revolution	**2nd revolution**	**3rd revolution**
19th century	20th century	21st century
Printing press	Radio, TV	Internet
Coal and Steam	Oil and Electricity	Renewable Energies

Table 1: Industrial revolutions and their drivers: communication and energy sources.

2. Making: FabLabs Today

Ever since their first inception in 2002[1], FabLabs equipped with digitally controlled machines that are available to ordinary people have started to spread a "coming revolution on [the] desktop" (Gershenfeld 2005): the revolution of personal digital manufacturing. As means of industrial production became easily accessible and designs were shared freely, hardware was likely to follow the evolution of Open Source software and that "a continuum from creators to consumers, servicing markets ranging from one to one billion" (Gershenfeld 2005, p. 21), would evolve, countering the paradigm of mass manufacturing and mass consumption with a peer-produced and community-based commons.

In practice, many disciplines have already experienced going Open Source and community-produced as the beginning of that Third Industrial Revolution. Not only was the software business fundamentally changed by the advent of Open Source software, but in music, "piracy is the new radio" (Young et al. 2012), in journalism, blogs and social media have taken much of the attention that printed papers used to get (Altermann 2008, Newman 2011), user-generated YouTube videos are displacing corporate news teams (PEJ 2012), and Wikipedia has outgrown printed encyclopaedias in volume, depth, recency and use (Okoli et al. 2012). However, the route to this new world of Open Source hardware and distributed manufacturing might be somewhat thornier than in software. There are at least two issues to be considered. First, it would be naïve to believe that Open Source software practices could be simply copied and applied to the manufacturing domain without any alteration

1 | Different authors determine the 'start' of FabLabs differently. I refer here to the presentation given by Bakhtiar Mikhak at the NSF site, visit on the 2nd of August 2002 (Mikhak 2002). The first FabLabs appear to have been opened that year as part of the outreach programme of Neil Gershenfeld's Center for Bits and Atoms.

or adaptation, ignoring the constraints and opportunities that the materiality of hardware entails. Second, more than two in three FabLabs are currently set up and run by institutions rooted in the 'old world order'. These institutions by their very nature are alien to lateral power relationships, struggle to understand polycentric structures and heterarchies, and fail to embrace a peer-production commons.

3. Open Source Hardware

Open Source hardware[2] is by no means a new phenomenon, and sharing of invention and product manufacturing information has been documented for the 18th and 19th century (e.g. Allen 1983, Nuvolari 2004, Bessen & Nuvolari 2012). James Bessen and Alessandro Nuvolari (2012) even conclude that "key technologies at the heart of industrialization [...] were, at times and places, developed through processes of collective invention" (Bessen & Nuvolari 2012, p. 12) and that "[i]n some cases [...] aggressive patenting put an end to a period of extensive knowledge sharing" (Bessen & Nuvolari 2012, p. 12). It is ironic that patenting has become blatant normality, to the extent that the number of patents filed in a country is used as the principal measure for its innovation performance, and that hardware has to be made 'open' again when it inherently and historically is open.

Yet, Open Source software practices cannot be simply copied and applied to the hardware domain; they have to be altered and adapted to account for the opportunities and constraints that the materiality of hardware entails.

First, there is inherent openness – hardware can be pretty self-explanatory about its composition. To keep that openness intact, the challenge lies in defeating the novelty requirement of related patent application or design registrations by open design techniques. Second, breaking up complex systems into simpler modules is not as common in hardware design as in software – despite being promoted as good design practice. Combining modules is potentially more complex than in software, as physical forces, mechanical fit and design considerations must be taken into account. Third, there are materials involved that may come at a cost and manufacturing processes that may not easily be accessed or require specialist tooling. Different strategies can be employed to overcome such barriers, such as using industrial side products as raw materials, pooling manufacturing resources, or using more universal fabricators. Fourth, the term hardware spans a much broader field than software and includes such nonrelated things as integrated circuits, home furniture and ship-to-shore container cranes. The different branches of hardware vary according to materials and technologies involved, manufacturing tools and processes, documentation customs and standards etc., hence, the aforementioned characteristics may apply to a different extent. Various initiatives have been started to define and certify Open Source hardware and relatedly Open (Source) design –

2 | Open Source Hardware is also a US registered service trademark owned by Fuhu, Inc. of El Segundo CA.

e.g. the TAPR radio amateur community[3], Open Collector[4], the Open Hardware project[5], the Open Source Hardware and Design Alliance, OHANDA[6], the Open Source Hardware User Group[7], the Open Hardware definition at Freedom Defined[8], the Open Source Hardware Logo[9], the Open Hardware Association[10], host of the annual Open Hardware Summit, and the Open Design working group of the Open Knowledge Foundation[11]. Online repositories of open hardware have been started to appear, too – e.g. Instructables[12], Thingiverse[13], or the Open Hardware Repository[14], to name but a few.

4. Peer-Produced Commons

Similar to Open Source software, this emerging ecosystem of Open Source hardware can be seen as a peer-produced commons – "thousands of volunteers [...] collaborat[ing] on a complex economic project" (Benkler 2002, p. 371) – as Yochai Benkler described it, a third model of production different from markets (that are organized by price signal) and firms (that are organized by hierarchical command and control). A peer-produced commons builds on lateral relationships. Peer production, according to Benkler (2002, p. 404), builds on four attributes of the Internet-based economy:

3 | For more information see the website, last viewed 25 September 2012 <http://tapr.org>, created 1993.

4 | For more information see the website, last viewed 25 September 2012 <http://opencollector.org>, created 2000.

5 | For more information see the website, last viewed 25 September 2012 <http://openhardware.org>, created 2002.

6 | For more information see the website, last viewed 25 September 2012 <http://ohanda.org>, created 2009.

7 | For more information see the website, last viewed 25 September 2012 <http://oshug.org>, created 2010.

8 | For more information see the website, last viewed 25 September 2012 <http://freedomdefined.org/OSHW>, created 2010.

9 | For more information see the website, last viewed 25 September 2012 <http://oshwlogo.com>, created 2011.

10 | For more information see the website, last viewed 25 September 2012 <http://oshwa.org>, created 2012.

11 | For more information see the website, last viewed 25 September 2012 <http://design.okfn.org>, created 2012.

12 | For more information see the website, last viewed 25 September 2012 <http://instructables.com>, created 2005.

13 | For more information see the website, last viewed 25 September 2012 <http://thingiverse.com>, created 2008.

14 | For more information see the website, last viewed 25 September 2012 <http://ohwr.org>, created 2009.

- Information is a non-rival good, it may be 'consumed' (used) by one consumer without preventing others to use it simultaneously. Obviously, this is also true for manufacturing information: hardware blueprints, manufacturing instructions, machine settings etc..
- Information can be produced at dramatically low cost. While producing physical goods will always incur the costs for materials, direct digital fabrication on community-owned machines – as, for instance, in FabLabs – decisively lowers manufacturing costs.
- Creative talent, the main human input to the process of creation, is best controlled by the creative individuals themselves as they "possess better information than anyone else about the suitability of their talents and their level of motivation and focus at a given moment to given production tasks" (Benkler 2002, p. 371).
- Information exchange and communication, key to the coordination of production processes, are cheap and efficient across the Internet (if used appropriately). Moreover, in distributed manufacturing, it is possible to create and distribute information globally and manufacture physical goods primarily locally or regionally, eventually reducing the amount of globally shipped goods.

Open Source hardware as a peer-produced commons might at least initially take different shapes in different economic contexts: "[T]he killer app for personal fabrication in the developed world is technology for a market of one, personal expression in technology [...] And the killer app for the rest of the planet is [to overcome] the instrumentation and the fabrication divide, people locally developing solutions to local problems" (Gershenfeld 2006).

Eric von Hippel, together with Jeroen de Jong and Stephen Flowers (2010), based his estimations on a representative study in the UK, concluding that consumers' annual product development expenditures were £5.1bn, which is 2.3 times the annual consumer product research and development expenditures of all firms in the UK combined (£2.2bn).

Such peer-production communities, including the FabLab community, are challenging some foundational assumptions about the free market. "What was formerly taken for granted or minimized in free-market theory – the role of social and civic factors in economic production – is becoming a powerful variable in its own right" (Bollier 2012, p. 35), as David Bollier states. Christian Siefkes (2008) seeks to generalize peer production 'into the physical world' and draws a picture of a society where peer production is the primary mode of production.

Yochai Benkler (2003) warns that, historically, structural economic patterns were determined within only a few decades after revolutionary technical developments and that "[t]he time to wake up and shape the pattern of freedom and justice in the new century is now" (Benkler 2003, p. 1276). "What decentralized and nonmarket information production generally, and peer production in particular, need, is a space free of the laws developed to support market- and hierarchy-based production" (Benkler 2003, p. 1273). For a "political economy of information" (Benkler 2003,

p. 1245), new ways are needed as to how to pursue autonomy, democracy, and social justice – the political sphere – and how to organize production and consumption – the economic sphere.

5. Institutional Embeddedness and Institutional Challenge

The second issue, organization of and governance in the FabLab community, deserves special attention, not only because peer-produced commons require special forms of governance, but also because there is an incongruence in the way makers as users of FabLabs and institutions as the main providers of FabLabs approach that issue.

On the one hand, makers in FabLabs are busy with their own manufacturing projects and make use of their lateral relations as needed, but do not normally bother about the organization of those relationships beyond those just-in-time needs. Occasionally, they wish for better, more effective access to resources in the network. So far, however, they have only come up with very few sustainable and scalable ideas to create new ways of organizing distributed personal manufacturing – organization and governance is not their core interest. Institutions, on the other hand, are rather concerned about organization, structures and governance, yet their solutions tend to be of conventional, hierarchical, top-down nature: centralized cathedral structures. Moreover, those solutions risk counteracting lateral approaches, suffocating emergent peer-to-peer initiatives – and they fail to get accepted by the makers.

Neil Gershenfeld points out that the power of the FabLab community is the bottom-up application of technology outside traditional institutions: "The message coming from the FabLab is that the other five billion people on the planet aren't just technical sinks, they are sources. The real opportunity is to harness the inventive power of the world to locally design and produce solutions to local problems. I thought that's a projection twenty years hence into the future, but it's where we are today. It breaks every organizational boundary we can think of. The hardest thing at this point is the social engineering and the organizational engineering, but it's here today" (Gershenfeld 2006).

In order to successfully develop the digital manufacturing ecosystem beyond a mere collection of individual tinkerers, a common understanding is needed of how such an ecosystem would function. Such a common understanding could build on a suitable theory, as there is nothing as practical as a good theory. However, canonical knowledge in business administration, industrial engineering and organization science on 'how to run a factory' and the collective wisdom of practitioners and consultants alike will only lead to hierarchies.

According to Elisabeth S. Clemens, "the imagery of the centralized, rationalized bureaucracy is increasingly unable to capture the empirical world confronted by organizational analysts" (Clemens 2005, p. 352), hence, insight needs to be found outside those disciplines. There is, indeed, a substantial body of knowledge about communities, collective action, self-organization and inverse infrastructures, as well as about peer-production and governing the commons.

6. Polycentric Systems

Communities, movements and collective action have been of research interest in social movement theory (see also the chapter 'The Movement' in this volume), and the topic has recently gained interest in organizational analysis and design (see e.g. Evans & Davis 2005). Siobhán O'Mahoney and Karim R. Lakhani (2011) discuss the impact of communities on organizations, concluding the following four key points:

- Communities help organizations emerge.
- Communities mediate the performance and growth of organizations.
- Communities can pose competitive threats to organizations.
- Communities outlive organizations.

In this sense, the FabLab community today is both threatening pre-existing organizations built around the provision of and education about technology and possibly helping new organizations emerge. Given the preference for lateral structures in the Third Industrial Revolution, those new organizations will develop polycentric or heterarchical forms as defined by David C. Stark, "with distinctive network properties [...] and multiple organizing principles" (Stark 2001, p. 71). A polycentric approach may be needed to solve the governance problems of the common-pool resources and the peer-produced commons of FabLabs. As Elinor Ostrom (2008) found, polycentric systems are one approach to solve collective-action problems related to the governance of public goods (commons) and common-pool resources.

Leonard Dobusch and Sigrid Quack (2010) compared the development of Wikipedia and Creative Commons in the years 2001 to 2008, finding that both started as relatively non-participatory, centralized organizations and developed into participatorier, decentralized structures. Creative Commons followed a strategy of decentralization first and participation later, while Wikipedia's strategy was based on participation first and was decentralized later. Mayo Fuster Morell (2011) found that Wikipedia also adapted organizationally to the changing needs of the community over time, and that it adopted a hybrid model for its infrastructure governance. The central Wikimedia Foundation adapted a traditional, representational democratic logic, while the community remained an innovative, elaborate, organizational model. This is approach is similar to the successful Open Source initiatives of the Linux kernel and the Apache http server development projects (Lanzara & Morner 2004).

These findings were also confirmed by recent research into 'inverse infrastructures', infrastructures that are formed bottom-up by means of many small private investments, in more than only the ICT sector (Egyedi & Mehos 2012). As Wim G. Vree (2003) points out, inverse infrastructures require different thinking at the administrative level: "The words 'design', 'construct' and 'implement' used in the classical approach could be replaced by 'bring about', 'cause to happen' or 'create optimum conditions for growth'" (Vree 2012, p. 276; his emphasis). Moreover, Tineke M. Egyedi (2012) portrayed inverse infrastructures as disruptive in the current institutional context and requiring more adaptive and robust infrastructure agreements, policies and regulation on national, regional and international level (Egyedi 2012, p. 259 et sqq.).

The policy recommendations proposed by the European Design Initiative (Thomson & Koskinen 2012) acknowledge this development: Design innovation in the 21st century, they argue, is characterized by social-based developments and collaboration in networks of designers and stakeholders, and Open Source design is based upon European values of diversity, low power distance and democracy (Thomson & Koskinen 2012, p. 38). Hence, they recommend to "create guidelines, codes of practice, legal frameworks and experimental spaces to promote the use of Open Design" (Thomson & Koskinen 2012, p. 45).

7. Making the Revolution

The topic of this chapter is the struggle for polycentric structures and a new peer-production commons in the FabLab community. But the chapter is also sub-headed 'Making the Third Industrial Revolution' – and taking this title literally, the chapter aims to provide more than just a definition of the problem. Therefore, I will subsequently address how to build effective forms of collective action and self-organization for the FabLab community and how to break free from traditional systems and creatively design new systems that tap into the capabilities of that community.

Elinor Ostrom and Charlotte Hess proposed to use the Institutional Analysis and Development (IAD) framework for analyzing knowledge commons (Hess & Ostrom 2007, p. 41). This primarily analytical framework comprises three clusters of variables that are rather broad:

1. The basic underlying factors or resource characteristics are the biophysical-technical characteristics, the attributes of the community and the rules-in-use (position of participants, boundary rules, authority, aggregation, scope, information availability, and pay-off rules),
2. The action arena made up by action situations and actors, and
3. The outcomes: patterns of interactions, outcomes, and the evaluation criteria that allow assessing these outcomes (see next section).

Beyond using this framework purely analytically, it can also serve as a guide for development. According to Ostrom and Hess, "the action arena [...] is an appropriate place to start when trying to think through the challenges of creating a new form of commons" (Ostrom & Hess 2007, p. 45). From the envisaged action situations, likely patterns of interaction and outcomes can be estimated. Physical and material conditions, community conditions, and rules-in-use that are likely to bring about those actions can then be derived.

In the case of FabLabs, the action arena would be a nested cluster of individual labs, regional networks, and the international community.

Action situations would include small groups or even individual users working on projects, sharing information, machining parts, as well as different individuals or groups coordinating machine access in a lab, or labs working together on projects or common infrastructures, such as an interconnected system for project documentation. Fab Academy, as well as the annual FabLab workshop and symposium, would be a specific action situation. Patterns of interaction would concern issues of overuse and underuse of resources, laser cutters likely being overused, and documentation repositories being underused. They would address free riding, productive and conflict behaviour, etc..

Possible outcomes would be cohesion or secession in the community, growth of reciprocity or conflict, recognition or ridicule. As basic underlying factors for FabLabs, one could identify the geographical and social location of labs, their users, user communities and institutional embedding, and implicit and explicit rules such as the FabCharter.

8. Measures of Success

Evidently, the evaluative criteria need to be established before starting an analysis and development exercise as sketched above – Elinor Ostrom and Charlotte Hess might have done this implicitly in their example when referring to 'creating a new form of commons'.

For the FabLabs as a peer-produced commons in the Third Industrial Revolution, those criteria would probably include (1) the protection of interests and creative freedom of makers, (2) wide access to new knowledge, processes and products, and (3) the extent to which it is possible to appropriately and effectively create and capture value.

According to these criteria, positive outcomes would be those that have a beneficial effect for makers: in a working peer-produced commons, they will have the authority to decide for themselves whether to contribute to the commons or not; they will be able to associate themselves and be associated with what they produce, they will be allowed to use and build upon what others have made; they will be able to build a reputation, establish productive relationships with peers, and economically sustain themselves.

For FabLabs, or whatever infrastructures the community would eventually use collectively, positive outcomes possibly could include the following: they may sustain their key enabling role for the community, they may be able to stay at the forefront of developments that further the development of the community and the spread of the Third Industrial Revolution, they may become hubs for transactions in the community, and the community may look after that collective infrastructure, also financially.

On the level of the community as a whole, positive outcomes could be in the internal and external workings of the community: Internally, the community would show cohesion without being sectarian, control of disruptive behaviour without the need for heavy policing, and diversity in any respect without the constant threat of falling apart. Externally, the community would be seen as one that not only provides access to direct digital manufacturing equipment to nerdy makers, but as one that also sparks innovation, empowers individuals and groups from all walks of life, and contributes to a thriving economy and the human condition in general.

9. A Roadmap

As Jeremy Rifkin points out, the Third Industrial Revolution will require "a wholesale reconfiguration of the economic infrastructure" (Rifkin 2012, p. 2) and "a massive retraining of workers on a scale matching the vocational and professional training at the onset of the First and Second Industrial Revolutions" (Rifkin 2012, p. 3). FabLabs can contribute to both the reconfiguration of the economic infrastructure and the (re)training of workers. FabLabs have the potential to tell a compelling story that can become part of the overall narrative of the Third Industrial Revolution, which Rifkin is missing in EU policy.

In order to be able to tell that compelling story, the FabLab community needs to stop being preoccupied with machines and sole practical making, and instead needs to wake up to the challenges of "the social engineering and the organizational engineering" (Gershenfeld 2006) and should start working on how to organize the ecosystem. That development could evolve around a roadmap of five key questions:

- How to build effective forms of collective action and self-organization for FabLabs?
- How to break free from traditional systems and creatively design new systems that tap into the capabilities of FabLabs?
- How to protect the interests and creative freedom of makers while also ensuring wide access to new knowledge, processes, and products?
- How to appropriately and effectively create and capture value?
- How to achieve equity and fairness?

It is crucial that the study of these questions and the development of answers are not left to the disciplinary professions of social scientists and management consultants. While their contribution can be crucial at times, any relevant development in a peer-to-peer community will have to come from within. Study will have to be participative, not purely observational. Design has to be emergent, not prescriptive. While theory can and should inform practice, practice also refines theory. Collaboration and a multiplicity of views are important, as is the question of how to evaluate development and monitor progress.

Making the Third Industrial Revolution is not an easy engineering or design task: learning and exploring the unknown will be required, a journey of continuous trial-and-error over several decades. In software, this is termed "perpetual beta" (O'Reilly 2005), which might loosely equal the notion of the 'learning organization' in management.

On this journey, we have to be prepared to get surprised, we must dare to fail, and we will have to disagree, but always in constructive ways.

References

Allen, RC 1983, 'Collective Invention', *Journal of Economic Behavior and Organization*, vol. 4, no. 1, pp. 1-24.

Altermann, E 2008, 'Out of print. The death and life of the American newspaper', *New Yorker*, viewed 20 September 2012, <http://www.newyorker.com/reporting/2008/03/31/080331fa_fact_alterman>.

Anderson, C 2012, *Makers: The New Industrial Revolution*, Crown Business, New York, USA.

Benkler, Y 2002, 'Coase's penguin, or Linux and the nature of the firm', *Yale Law Journal*, vol. 112, pp. 369-446, viewed 20 September 2012, <http://www.yalelawjournal.org/images/pdfs/354.pdf>.

Benkler, Y 2003, 'Freedom in the commons: Towards a political economy of information', *Duke Law Journal*, vol. 52, pp. 1245-1276, viewed 20 September 2012, <http://scholarship.law.duke.edu/dlj/vol52/iss6/3/>.

Bessen, JE & Nuvolari, A 2012, 'Knowledge sharing among inventors: Some historical perspectives', in Harhoff, D & Lakhani, K (eds.), *Revolutionizing innovation: Users, communities and open innovation*, MIT Press, Cambridge. Pre-Print: Boston Univ. School of Law, Law and Economics Research Paper No. 11-51; LEM Working Paper 2011/21. Available at <http://www.bu.edu/law/faculty/scholarship/workingpapers/documents/BessenJ-NuvolariA101411fin.pdf>.

Bollier, D 2012 'The Growth of the Commons Paradigm', in Hess, C & Ostrom, E (eds), *Understanding Knowledge as a Commons*, MIT Press, Cambridge, MA.

Clemens, ES 2005, 'Two kinds of stuff: The current encounter of social movements and organizations', in Davis GF, McAdam D, Scott WR & Zald MN (eds.), *Social movements and organization theory*, Cambridge University Press, New York, pp. 351-366.

Dobusch, L & Quack, S 2010, 'Managing boundaries between organizations and communities: Comparing creative commons and Wikimedia', paper presented at the 3rd Free Culture Research Conference, Berlin, October 8-9, 2010, viewed 20 September 2012, <http://wikis.fu-berlin.de/download/attachments/59080767/Dobusch-Quack-Paper.pdf>.

Egyedi, TM & Mehos, DC (eds.) 2012, *Inverse infrastructures. disrupting networks from below*, Edgar Elgar, Cheltenham.

Evans WR & Davis WD 2005, 'High-performance work systems and organizational performance: The mediating role of internal social structure', *Journal of Management*, no. 31, pp. 758-775.

Fuster Morell, M 2011,'The Wikimedia foundation and the governance of Wikipedia's infrastructure. Historical trajectories and its hybrid character', in Lovink, G & Tkacz, N (eds.), *Critical point of view. A Wikipedia reader*, Institute of Network Cultures, Amsterdam. pp. 325-341.

Gershenfeld, N 2005, *Fab. The coming revolution on your desktop – from personal computers to personal fabrication*, Basic Books, New York.

Gershenfeld, N 2006, FabLabs, TEDTalk, viewed, 2 January 2013, <http://blog.ted.com/2009/08/11/fab_labs_neil_g/>.

Von Hippel, E, de Jong, JPJ & Flowers, S n.d., *Comparing business and household sector innovation in consumer products: Findings from a representative study in the UK*, viewed 20 September 2012, <http://ssrn.com/abstract=1683503>.

Lewin, K 1952, *Field theory in social science: Selected theoretical papers by Kurt Lewin*, Tavistock, London.

Lanzara, GF & Morner, M 2004, 'Making and sharing at electronic crossroads: The evolutionary ecology of open source', paper presented at the The Fifth European Conference on Organizational Knowledge, Learning, and Capabilities, University of Innsbruck, Austria, 2-3 April 2004, viewed 20 September 2012, <http://www2.warwick.ac.uk/fac/soc/wbs/conf/olkc/archive/oklc5/papers/j-3_lanzara.pdf>.

Newman, N 2011, Mainstream media and the distribution of news in the age of social discovery. How social media are changing the production, distribution and discovery of news and further disrupting the business models of mainstream media companies, Reuters Institute for the Study of Journalism, Oxford, viewed 20 September 2012, <http://reutersinstitute.politics.ox.ac.uk/fileadmin/documents/Publications/Working_Papers/Mainstream_media_and_the_distribution_of_news_.pdf>.

Nuvolari, A 2004, 'Collective invention during the British industrial revolution: The case of the Cornish pumping engine', *Cambridge Journal of Economics*, vol. 28, no. 3, pp. 347-363.

O'Reilly, T 2005, *What Is Web 2.0* , viewed 20 September 2012, <http://oreilly.com/web2/archive/what-is-web-20.html>.

O'Mahoney, S & Lakhani, KR 2011, 'Organizations in the shadow of communities', in Marquis, C, Lounsbury, M & Greenwood, R (eds.), *Communities and organizations*, Emerald Group, Bingley, pp. 3-36.

Okoli, C, Mehdi, M, Mesgari, M, Nielsen, FA & Lanamäki, A 2012, *The people's encyclopedia under the gaze of the sages: A systematic review of scholarly research on Wikipedia*, viewed 20 September 2012, <http://ssrn.com/abstract=2021326 or http://dx.doi.org/10.2139/ssrn.2021326>.

Ostrom, E 2008, *Polycentric Systems as One Approach for Solving Collective-Action Problems*, viewed 20 September 2012, <http://ssrn.com/abstract=1304697>.

Ostrom, E & Hess, C 2007, 'A Framework for Analyzing the Knowledge Commons', in Hess, C & Ostrom, E (eds.), *Understanding Knowledge as a Commons*, MIT Press, Cambridge, MA.

PEJ 2012, 'YouTube & the news: A new kind of visual journalism', *PEJ report*, Pew Research Center, Project for Excellence in Journalism, 16 July 2012, viewed 20 September 2012, <http://www.journalism.org/node/30171>.

Rifkin, J 2012, 'Beyond austerity: A sustainable third industrial revolution economic growth plan for the European Union', an executive summary of Jeremy Rifkin's keynote speech for the Mission Growth Summit: Europe at the lead of the new industrial revolution, hosted by The European Commission, May 29th 2012, pp. 1-4, viewed 20 September 2012, <http://ec.europa.eu/enterprise/policies/innovation/files/mg-speech-rifkin_en.pdf>.

Rifkin, J 2011, *The Third Industrial Revolution. How Lateral Power is Transforming Energy the Economy and the World*, Palgrave Macmillan, New York.

Siefkes, C 2008, *From exchange to contributions. Generalizing peer production into the physical world*, Siefkes, Berlin, viewed 20 September 2012, <http://peerconomy.org/text/peer-economy.pdf>.

Stark, DC 2001, 'Ambiguous assets for uncertain environments: Heterarchy in postsocialist firms', in DiMaggio,P (ed.), *The twenty-first-century firm: Changing economic organization in international perspective*, University Press, Princeton, pp. 69-104.

Thomson, M & Koskinen, T (eds.) 2012, *Design for growth and prosperity. Report and recommendations of the European design leadership board*, DG Enterprise and Industry of the European Commission, Brussels, viewed 20 September 2012, <http://ec.europa.eu/enterprise/policies/innovation/files/design/design-for-growth-and-prosperity-report_en.pdf>.

Vree, WG 2003, 'Internet en Rijkswaterstaat: een ICT-infrastructuur langs water en wegen', Inaugurele rede uitgesproken bij de aanvaarding van het ambt van hoogleraar ICT aan de faculteit Techniek, Bestuur en Management van de Technische Universiteit Delft, 29. Jan. 2003, viewed 20 September 2012, <http://www.tbm.tudelft.nl/en/about-faculty/departments/iss-department/ict-section/staff/wim-vree/publications/inaugural-speech/>.

Vree, WG 2012, 'The Internet and Rijkswaterstaat: An ICT infrastructure for roads and waterways', in Egyedi, TM & Mehos, DC (eds.), *Inverse infrastructures. Disrupting networks from below*, Edgar Elgar, Cheltenham, pp. 267-290.

Young, N, Mossberg, W & Kafka, P 2012, *Dive into Media: Neil Young*, viewed 20 September 2012, <http://allthingsd.com/video/?video_id=1598C8DC-7B17-4E42-A95A-DE703ACC12A9>.

COMMUNITY AND ENVIRONMENT

NOTES ON COMMUNITY AND ENVIRONMENT

BART BAKKER

My name is Bart Bakker, and my job is technology assessment: that means I have to think about the expectations, possibilities and limitations of emerging technology. The secret of this job is to look beyond the 'new technology hype'. It also means that it can be hard to convince people sometimes, as everyone already has a certain assumption about technology. Back in 1993, for example, at a time where there were hardly any mobile phones, I spotted the Internet as a huge potential and disruptive technology. Hence, in 1994, I secured one of the 68 .nl Internet domains in the Netherlands at this time, despite my boss' disbelief in this new technology. However, being convinced by the Internet's potential, I cofounded Waag Society, an institute for art, science and technology in Amsterdam – that also hosts the FabLab in Amsterdam.

To explain how I first got in touch with FabLabs means to go back to the year 2007, when a small FabLab was installed during the art project 'El Hema' in Amsterdam. I was convinced that this was a new, disruptive type of technology, which was about to emerge, so I was eager to get involved. A month later, after having built a 3D printer at Picnic07 (a media festival), I was ultimately taken by the FabLab idea. Hence, I am now involved with Protospace, the main FabLab in Utrecht, and with the Dutch FabLab Society.

My idea to set up a mini FabLab slowly 'developed', it was not planned from the beginning. At Protospace, there were people queuing in front of the laser cutter, waiting for their turn. Most of the time, they were waiting just to do a small test. And then they had to wait another half an hour – or more. Having watched this scenario, I got interested in small 'cheap chinese laser cutters' for extra capacity; sadly, their user software – though mechanically quite good – was of low standard, hence it was not fit for a FabLab. So I kept the little cutter for experimenting with it.

Moreover, I had built an Ultimaker, as well as a Mantis router. It was then that I realized I had a mini FabLab in my own garage (and online: www.minifablab.nl), so I invited people to come and make mainly small things.

And people did come, including artists, jewelry makers, and model makers, but to my surprise they did not want to make things, they wanted to discuss small laser cutters in order to buy one for themselves. This is how the scope of the mini FabLab shifted from a small maker space to a place to get advice on small laser cutters and gantry routers, as well as on how to start a FabLab. Even though this was never the idea behind it, there proved to be a demand for it. (However, I am currently also assisting in setting up a full-blown mid-sized FabLab in just two months' time, which I enjoy quite a lot.)

I was very dissatisfied with the handling of the small laser cutters though. While there is a huge demand for desktop machines that can cut A4 or A3, there is no Open Source software, only the user-unfriendly Chinese programs. Therefore, we researched the possibilities of an Open Source project to provide both a controller card and the software to drive such machines. However, that is not a simple task, otherwise it would have been done already. We initiated LaOS, Laser cutter Open Source, and it took off. Now, one can simply hit 'Ctrl-P' and the laser cutter cuts your job from Inkscape, on any of the platforms. The controller board is available now and the software is available in a beta-version. The best thing that could happen is that the Chinese copy the LaOS board – it is open – and offers it in their small machines. That would make small laser cutters affordable for everyone. After all, the laser cutter is the driving force of any FabLab.

Judging from experience, there are a few technical challenges for FabLabs. The biggest challenge is probably education, that is, how to teach the use of the digital machines. If you cannot make a drawing in a 2D or 3D program, a FabLab has little to offer you. Therefore, the main challenge for any FabLab is to help people learn to use the machines – this might be a non-issue 20 years from now, but we have to bridge those 20 years. However, mastering the technology is not too difficult. In FabLabs, we should lower the threshold not only by offering courses, but also by simplifying access to the machines and making user interfaces more facile.

Looking into the future, I think most labs will become smaller than the current ones, physically more of a garage/schoolroom size. Yet, I predict they will be strongly interconnected, not only the people, but also the machines and the underlying repositories of drawings and designs. The current machine control is cumbersome and so are the current 3D design tools. But my presumption is that this will change in five years' time. I also expect the FabLab community, still based on the MIT paradigm, to shift from machine orientation to usability and easy access, as well as to more interconnection. A FabLab will no longer be seen as a lab, but just as a place to easily make things. Commercial labs will arise, just like the photocopy shops did in the 1970s, but their function will probably proliferate to private homes. I therefore think we should embrace a vision of a myriad of small and interconnected FabLabs and be open to help that community grow. As for 3D printing, I assume that it is going to be huge. But the way to go and the technical processes are not yet determined. We can compare the process to that of microcomputers in the 1980s. At that time, a black-and-white daisywheel printer was a revolution compared to the matrix printer. Now they are both gone and you get a faster and better printer for free if you buy some ink cartridges. This will also happen with fabbing machines. At this moment, the 3D printers are slow, the results crude and the methods clumsy. In as short as five years' time, Fused Deposition Modeling-based 3D printers like the Ultimaker and the Replicator may have to yield to a total different approach. But what will grow are the wish and the ability to make one's own thing, which has been shown by the current 3D printers. I am sure localized production will disrupt more than we can imagine today.

DIGITAL FABRICATION AND 'MAKING' IN EDUCATION

THE DEMOCRATIZATION OF INVENTION

PAULO BLIKSTEIN

1. Digital Fabrication: Logo for Atoms

A quote often attributed to Seymour Papert states that if a teacher from the 16th century would time-travel to the present, he or she would have no problem entering a school and teaching a class. Historical documents from that time show that he could not be more accurate. The Treviso Arithmetic from 1478 teaches students how to do multiplication and division using 'exactly' the same paper-based algorithms we use today. Several descriptions of 16th century schools and their curricula look strikingly similar to today's mathematics classes, such as a well-known school in Florence run by Master Francesco Ghaligai in 1519 which had a "heavy emphasis on memorization [...] and [...] procedures" (Swetz & Smith 1987, p. 23) and a curriculum comprised of units on multiplication, practice in the use of algorithms, division, fractions, and the rule of three (Swetz & Smith 1987).

A thriving 15th century Venice saw the appearance of the Treviso Arithmetic in a time of great need for a new type of mathematics. The Indo-Arabic system was proving itself to be faster and more practical than the abacus, and soon Venetians realized that it was also easier to learn. All the pieces were falling into place: a new set of societal needs, new technologies, new ways of using knowledge, and the recognition that a task previously monopolized by experts was potentially accessible to the masses "restructurations" (Wilensky & Papert 2010).

Every few decades or centuries, a new set of skills and intellectual activities become crucial for work, conviviality, and citizenship – often democratizing tasks and skills previously only accessible to experts. Fast forward to the early seventies: computer programming was becoming one of those new activities (Papert 1991). But computers in those years were large, expensive, awkward machines, and the idea of using them as a medium for personal expression and learning was inconceivable – in the same way that the abacus establishment derided the Treviso techniques. The educational establishment put down the idea of programming as a fundamental pedagogical goal: it was too difficult for children to learn, and unlikely to improve learning in math and science.

However, since the seventies, researchers have been hard at work creating tools to make programming easier to learn. Programming tools such as Scratch (Resnick et al. 2009) and NetLogo (Wilensky 1999) have achieved unprecedented popularity and made coding accessible to millions of students and teachers. The world caught up with the idea that not only computational media could be a vehicle for powerful ideas in mathematics, engineering, and science – an important new kind of literacy – but it was an approachable activity in schools.

Digital fabrication and making could be a new and major chapter in this process of bringing powerful ideas, literacies, and expressive tools to children. Today, the range of accepted disciplinary knowledge has expanded to include not only programming, but also engineering and design (Astrachan et al. 2009; Yasar & Landau 2003). In addition, there are calls everywhere for educational approaches that foster creativity and inventiveness.

The analogy with the development of Logo is clear: simultaneously, digital fabrication technology became better and more accessible, and the intellectual activities enabled by the new technology became more valued and important. What Logo did for geometry and programming – bringing complex mathematics within the reach of schoolchildren – fabrication labs can do for design and engineering. Digital fabrication is Logo for atoms.

In this chapter, I first briefly review the history of engineering education to show the rise and fall then rise again of the making and building as curricular foci. I then discuss the theoretical underpinnings of project-based, student-centered, constructionist learning, showing that much of what digital fabrication labs can enact was already predicted and advocated in the theories and writings of John Dewey, Seymour Papert, and Paulo Freire. The following section approaches the educational benefits of digital fabrication and how it could be a unique tool in the hands of progressive educators. In the final part of the chapter I present not only four prototypical episodes that exemplify the advantages and perils of FabLabs in schools, but also some guidelines for the design of learning environments incorporating these types of technologies.

1.1 Technological Knowledge: From Skills to Literacy

In 1999, the National Research Council issued a landmark report stating that technology was changing too fast for the 'skill-based' approach to be effective and instead called for a 'fluency' approach. They suggested technological education to include the development of adaptive, foundational skills in technology and computation, in particular intellectual "[c]apabilities [to] empower people to manipulate the medium to their advantage and to handle unintended and unexpected problems when they arise" (National Research Council 1999, p. 2-3).

The same concerns were echoed in the later report, Technically Speaking: Why All Americans Need to Know More About Technology (National Research Council 2002), which confirmed the demise of the 'computer skills approach' and recognized that decades had been lost teaching dated skills to millions of students. It called for a move from 'computer skills' towards 'computational fluency' or 'literacy' (diSessa 2000) and broadening "technological literacy" (National Research Council 2002) to include basic engineering knowledge, and the nature and limitations of the engineering design process. (National Research Council 2002)

The report also introduced an important distinction, which resonated with the concerns of educational theorists such as Seymour Papert and Andrea diSessa: the recognition of a difference between 'technological literacy' (a general set of skills and intellectual dispositions for all citizens) and 'technical competence' (in-depth knowledge that professional engineers and scientists need to know to perform their work). The distinction identifies fluency with technology as no longer a vocational skill or a way to train future science, technology, engineering and mathematics (STEM) workers, but knowledge valuable for every citizen. Since the publication of the 2002 Technically Speaking report, several other developments in research, technology and policy have further supported this need: the acceleration of technological innovation, further automation of routine jobs (Levy & Murnane 2004), ubiquity of Open Source hardware and software, and the development of low-cost digital fabrication tools (Gershenfeld 2007). These national reports and societal developments are noteworthy because they signal the mainstream acceptance of Papert's once controversial vision. Taken together, the once dismissed idea of children programming computers was not only embraced, but developed into a much larger vision of students participating in sophisticated activities that were previously restricted to specialized professionals, such as robotics, environmental sensing, data analysis, advanced science, and engineering design.

1.2 The Demise of the Shop Class and the Rise of the Digital Fabrication Lab

Notwithstanding the natural content overlaps amongst science and engineering disciplines, they are fundamentally different. While a scientific investigation is typically concerned with finding the one law to explain many natural phenomena, a technological investigation typically finds many solutions for the same problem (Atkin 1990). A typical school science lab is designed for rigorous, disciplined, and scripted experiences in which students are guided towards the re-discovery of a unifying principle. School science labs are architected to facilitate and optimize such a process – but would those spaces be appropriate for engineering and design?

Despite engineers' dependence on basic scientific knowledge to do their work, their epistemology even precedes science; humans have been creating tools and altering their environment much before the inception of the scientific method. In fact, engineers' 'ethos' as inventors and tinkerers, in both K-12 and college education, survived up to the fifties and sixties, after which there was a significant push towards analysis and mathematics, and away from traditional 'shop work' (Grinter 1955), which was overwhelmingly present in curricula during the first half of the 20th century (Dym 1999). The 'professional engineer' of the first half of the 20th century was replaced by the 'scientific engineer' of the second half (Tryggvason & Apelian 2006), mostly motivated by the end of the abundant Apollo-era funding – less expensive theoretical classes prevailed over engineering labs or design work (Feisel & Rosa 2005). Over time, this resulted in the removal of the engineering design experience from not only college curriculum, but also from K-12 education. Shop class became 'vocational education' for those who supposedly could not handle 'serious' math or science.

Two independent processes started to reverse this trend. First, around the eighties, faculty and employers started to feel that the design-deprived engineering graduates were not well prepared do to any real engineering design work, which had started to become more important (Sheppard & Jenison 1997). Second, in the early 2000s, prototyping equipment, such as laser cutters and 3D printers, dramatically dropped in price, and Open Source hardware further popularized these technologies. Suddenly, corporate product development moved towards a 'studio' model in which groups of engineers and industrial designers could create prototypes in days instead of months: consequently the nature of product engineering was radically transformed. Gershenfeld and colleagues (Gershenfeld 2007; Mikhak et al. 2002) at MIT (Massachusetts Institute of Technology) were the first to package such equipment in a standardized low-cost lab and deploy it in both community centers and universities around the globe: the FabLab was born. Gershenfeld's network of FabLabs quickly spread in all five continents, and spurred a vibrant global movement. Four years later, the Make: Magazine, a monthly publication dedicated to DIY enthusiasts and tinkerers was created, and the Maker Faire, a large science and engineering fair in California, launched with great success.

2. Dewey, Papert, and Freire: Theoretical Pillars for Digital Fabrication and 'Making' in Education

Toward the end of the 2000s, researchers and educators started to consider the use of digital fabrication in education. In 2008 Stanford University launched the FabLab@School project, and started building FabLabs in K-12 schools around the world. In 2009 the MC2STEM High School in Ohio (USA) opened its first digital fabrication lab. In 2011 the Maker Media launched the MakerSpace project with DARPA (Defense Advanced Research Projects Agency, an US military research agency) funding. In 2011 and 2012 alone countless museums, schools, community centers, and libraries announced plans to build digital fabrication and making facilities – it became mainstream. Despite this resurgence of fabrication labs and making in formal and informal settings, the ideas behind this movement are at least a century old. Digital fabrication and making are based on three theoretical and pedagogical pillars: experiential education, constructionism, and critical pedagogy.

Since Rousseau's invention of childhood (Rousseau 1961), progressive education theorists have been questioning the prevalent assumptions of their time regarding the project of education, and have been prescribing more experiential, student-centered approaches. The idea that education should be more experiential and connected to real-world objects is originally attributed to John Dewey but also to many other scholars and innovators (Dewey 1902; Freudenthal 1973; Fröbel & Hailmann 1901; Montessori 1964, 1965; von Glasersfeld 1984).

Critical pedagogy scholars, Freire in particular, criticized school's "banking education" (Freire 1974, p. 57) approach and the decontextualization of curriculum (Freire 1974; Illich 1970). Freire introduced the idea of culturally meaningful curriculum construction, in which designers get inspiration from the local culture toward creating 'generative themes' with members of these cultures. Freire was also an advocate for education as a form of empowerment, and argued that learners should go from the 'consciousness of the real' to the 'consciousness of the possible' as they perceive the 'viable new alternatives' beyond 'limiting-situations' (Freire 1974). Therefore, students' projects should be deeply connected with meaningful problems, either at a personal or community level, and designing solutions to those problems would become both educational and empowering (Blikstein 2008; Cavallo 2000).

Seymour Papert shares Paulo Freire's enthusiasm for unleashing the latent learning potential of students by providing environments in which their passions and interests thrive. A mathematician by training, who then worked with Jean Piaget for many years, Papert pioneered the use of digital technologies in education (Papert 1987). Yet Papert's reasons for advocating the use of computers in education are far from technocentric – some of his motivations are very similar to Freire's. Papert's Constructionism builds upon Piaget's Constructivism and claims that the construction of knowledge happens remarkably well when students build, make, and publicly share objects. His theory is at the very core of what making and digital fabrication mean for education, and underlie what many enthusiasts of the maker movement propose – even if many are not aware of it. Papert's words describe precisely the relationship between making and learning: "Construction that takes place 'in the head' often happens especially felicitously when it is supported by construction of a more public sort 'in the world' – a sand castle or a cake, a Lego house or a corporation, a computer program, a poem, or a theory of the universe. Part of what I mean by 'in the world' is that the product can be shown, discussed, examined, probed, and admired [...] It attaches special importance to the role of constructions in the world as a support for those in the head, thereby becoming less of a purely mentalist doctrine" (Papert 1980, pp. 142-143; his emphasis).

Papert advocates technology in schools not as a way to optimize traditional education, but rather as an emancipatory tool that puts the most powerful construction materials in the hands of children – again, another idea that inspired the resurgence of the maker sensibilities. These protean machines, which would enable students to design, engineer, and construct, would cater to many forms of working, expressing, and building. This chameleonesque adaptivity, which is embedded in technology, permits the acknowledgement and embracing of different learning styles and epistemologies, engendering a convivial environment in which students can concretize their ideas and projects with intense personal engagement. In a typical constructionist learning environment, there is rarely a fixed curriculum. Children use technology to build projects, and teachers act as facilitators of the process.

The Logo programming language was the first attempt in education to demonstrate that the computer is not only an information and communication device, but also an expressive tool for construction and self-expression. In the early nineties, Papert extended the powerful ideas of Logo to the physical world by making robotics accessible to children through the Lego Mindstorms kit and the Cricket (Martin & Resnick 1993; Martin 1994). More recently, new developments are putting cutting-edge hardware and software in the hands of children to conduct advanced scientific explorations (Blikstein 2010; Blikstein et al. 2012), create interactive textiles (Buechley 2006; Buechley & Eisenberg 2008; Buechley et al. 2008), build electronic jewelry (Sylvan 2005; Perner-Wilson et al. 2011), design participatory simulations and games (Wilensky & Stroup 1999), program videogames (Millner & Resnick 2005; Sipitakiat et al. 2004), design virtual robotic systems (Berland & Wilensky 2006; Berland 2008), create sophisticated 3D worlds and games through programming (Cooper et al. 2000), build new types of cybernetic creatures (Schweikardt & Gross 2006; Raffle et al. 2004) and explore environmental science and geographical information systems (Edelson 2000).

These toolkits and technologies prepared the ground for the popularity of the maker movement and digital fabrication. They showed that it was possible to engage children in complex uses of technology, that those same children could actively construct with technology rather than just consume technological products. They revealed how the ideas and intellectual passions of children could be powerful and generative, and that the perceived difficulties of many of those tasks were due to deficient design rather than learners' cognitive deficiencies. Rather than random developments, these new technologies, materials, and toolkits were deeply influenced by the theoretical constructs put forth by Dewey, Papert, and Freire around constructive uses of technology, culturally-aware education, experiential learning, and interest-driven curricula.

3. Why Do We Need Digital Fabrication Labs in Schools?

The plethora of constructionist toolkits created and deployed in the 2000s, with improved and friendlier designs, coincided with the development of the FabLab concept by Neil Gershelfeld at MIT and the popularity of the Maker Faire – the perfect storm was in place. At that time, after having conducted tens of robotics and invention workshops in schools, I was disappointed by the fact that students did not have a place to continue and deepen their projects – and projects would die after the workshop or the final expo. Schools manifest how they value a particular activity by building a space for it. If sports are important, schools build a gym and a basketball court. If music education is in demand, schools set up music rooms. Only then can likeminded students gather together, hang out, do projects, talk about them, and create a productive subculture in schools. Unfortunately, I realized that there was no such space for engineering and invention. Even when schools had robotics labs, they were highly gender-biased and not inviting for most students. Robotics labs and science labs were not disruptive spaces anymore. Therefore in 2008 I started to work with schools around the world to establish digital fabrication labs – the FabLab@School project was born.

I realized that digital fabrication had the potential to be the ultimate construction kit, a disruptive place in schools where students could safely make, build, and share their creations. I designed those spaces to be inviting and gender-neutral, in order to attract both the high-end engineering types, but also students who just wanted to try a project with technology, or enhance something that they were already doing with digital fabrication.

Both programming and educational robotics enhanced an existing activity with a powerful new expressive medium. Logo programming reinvented differential geometry by adding computer algorithms to children's everyday bodily movements like 'forward', 'turn right', 'turn left'. Robotics kits added computational behaviors to familiar materials (crafts, Legos, wheels) and behaviors ('light up if dark', 'bounce off the walls', 'follow the dark line'). Each one of them made possible for new forms of expressiveness by adding a carefully designed technological layer to everyday, familiar materials and practices.

Digital fabrication is a new chapter in this story. Especially in low-income schools, students would often tell me that they used to make and build things with their parents and friends, and often had jobs in garages, construction companies, or carpentry shops. However that experience was disconnected from their school life, since they did not see a link between the intellectual work in the classroom and the manual labor in the wood shop. Because of bias inherit within the educational system their own forms of engineering and tinkering, stripped down of any form of mathematical of scientific content, were looked down upon by society and by themselves.

Enhancing existing practices and expertise. One of the first and most striking results of the initial workshops in digital fabrication is that students reported they have gained a new appreciation for the 'manual' labor they used to do, and also for the occupation of their parents. In the lab, students had to first design their creations on a computer, often after several types of measurements and calculations. However, they were still constructing, building, and using their hands, but all the work was permeated with two socially valued practices: computation and mathematics. Again, the familiar practices of building and making were augmented with computational tools, which generated not only more refined and sophisticated projects, but also empowerment and increased self-esteem. This proved to be a crucial Freirean principle for the design of digital fabrication experiences. By building onto students' familiar practices and adding a layer of expressive technologies, a digital fabrication lab, which merges computation, tinkering and engineering, has the potential to augment rather than replace familiar and powerful practices that students already possess, therefore they can recognize their own previous expertise in what they accomplish in the lab, rather than acquiring a new identity altogether.

Accelerate invention and design cycles. An additional benefit of digital fabrication is that it accelerates the processes of ideation and invention. It eliminates manual dexterity as the 'middleman' in transforming an idea into a product, so students can focus their attention on improving the design rather than taking care of mundane issues with the materials – and many more cycles of redesign are possible in the same time interval. Moreover, as I consistently observed, the fact that the products generated in the laser cutter and the 3D printer were aesthetically pleasing had a strong impact in students' self-esteem – instead of taking home asymmetric and fragile cardboard prototypes, they were building functional 3D objects with a near-professional finish – it wasn't 'school stuff', it was the 'real thing'.

Long-term projects and deep collaboration. We also observed that the establishment of this new space in schools allowed students to engage in intellectual activities and practices that would not be possible anywhere else, and experience new ways of work and novel levels of team collaboration. A real engineering project takes several cycles of design and redesign. It does not fit the one-size-fits all 50-minute format. The digital fabrication lab provided a 'safe space' for long-term projects, which in turn enabled students to face (alone or in groups) a new and intense experience: failure. Learning how to manage failure – something rarely taught in schools – ended up being another crucial educational benefit of the lab work. As we will see in many of the vignettes, through several cycles of failure and redesign, students not only achieved incredibly original and complex designs, but also became more persistent, learned to work in heterogeneous teams, and became better at managing intellectual diversity.

4. Four Vignettes and Many Lessons About Digital Fabrication in Education

In the following four vignettes, I discuss positive and negative scenarios of the implementation of digital fabrication in education based on the categories I just described. I exemplify some of the learning outcomes, and offer recommendations for the design and management of such spaces. Each vignette illustrates one or two important principles, and in particular I discuss (a) the dangers of trivialization, (b) the potential for deep engagement in projects of unprecedented complexity, (c) the power of interdisciplinary work, (d) contextualized learning in STEM, and (e) intellectualization and re-evaluation of familiar practices.

4.1 The 'Keychain Syndrome', or the Temptations of Trivialization

For the first digital fabrication workshops we held in 2009, I designed introductory activities to get students acquainted with the machines: semi-structured short projects such as creating a keychain, a nametag, or an acrylic sign for a sports team. On a technical level these projects required students to learn how to cut and engrave using the laser cutter, use vector drawing software to create and combine geometric shapes, and import/manipulate bitmapped images from the web.

I assumed that by asking them to create highly-personal objects, such as keychains and nametags, students would get excited about the technologies not only because they would create objects for everyday use, but they would decorate their rooms, school materials, and clothes with them, attracting the attention of family members and other students in the school. They would feel proud of their creations and associate their newly acquired engineering skill to the production of socially valued artifacts.

Students engaged with enthusiasm in the creation of their keychains. The plan worked. For the second session, they came back even more excited about their objects – parents, friends, even teachers wanted an acrylic keychain. Students lined up by the laser cutter to make more keychains. Excitement was in the air. Digital fabrication was succeeding, and students – both girls and boys – were very excited about 'making stuff'.

By the third session, my team had decided that it was time to move on to new activities – in particular, I wanted to introduce robotics and electronics. I rounded up students at the beginning of the session and ran a short robotics tutorial, teaching them how to hook up sensors and motors, and write simple programs. At the end of the workshop, some students came to talk to me and asked permission to use the laser cutter for some new keychains. I postponed robotics for another day. By the fourth session, I knew something was wrong. The workshop became a keychain factory, and students would not engage in anything else. The plan worked too well – it backfired. Students found an activity that was personally meaningful, produced professional looking products that were admired and envied, and used a high-tech device. However, as much as it was a very effective solution to engage them in digital fabrication, it offered a too big reward for a relatively small effort, to produce an object that did not include any computation or complex constructive challenges. Ironically, it is as if students had discovered exactly what manufacturing is about – mass-producing with little effort – and were making the best of it. Students 'cracked' digital fabrication and were using the lab as a fabrication facility, rather than a place for invention.

The following dialogue, which took place several days into the workshop, illustrates the seductions of the 'keychain syndrome':

Facilitator: What would you do if you had a laser cutter at home?

Megan: I would make keychains.

Nancy: Yeah, and sell 'em.

Facilitator: Keychains? What kind?

Megan: Like, these (she takes out a collection of keychains that she had recently printed).

Facilitator: Anything else?

Megan: No, just keychains.

But there was a more systemic issue at play – friends and family were focusing on the only values that they know, not coincidentally values which schools have traditionally focused on: valuing product over process. In that sense, digital fabrication is a type of Trojan horse: it introduces in schools a 'genre' of tools that have the very special property of easily generating aesthetically pleasing, almost magical products. Therefore, for the student-creator, there is a conflicting incentive: (i) obfuscate the simplicity of the process ("I used this laser cutter machine, it's science fiction, it's really complicated"), and enhance the value of the product to others, or (ii) make the process transparent ("I used the laser cutter, it's actually not so hard to do keychains, the machine did most of the work!"), and reveal the triviality of the product.

For the educational designer and facilitator, it is fundamental to understand this incentive system to avoid this potentially harmful aspect of this 'genre' of machines. The feedback loop that the first incentive (obfuscating the simplicity of production) generates is that students get engrossed in the production of the same type of simple products. In the case of the second incentive, students are led to 'un-trivialize' the product given the new level of product complexity that digital fabrication enables them to achieve. In the first case, despite appearances, we 'schoolify' and trivialize the lab, in the second, we make it a place for excellence and inquiry. The solution, however, is not inconsequential – while the product-over-process conundrum does not resolve itself, there will always be an incentive for simple, well-polished products, as opposed to messy, complex, and potentially 'ugly' projects. Unless educational designers unveil the real incentive systems at play in the classroom, teachers who reward students based on quick completion times, quality of solution, and efficiency, might actually be fostering classrooms in which students rarely venture outside of what they already know (Abrahamson, Blikstein & Wilensky 2007).

The 'keychain syndrome', therefore, revealed two of the crucial elements of learning environments based on digital fabrication. First, the equipment is capable of easily generating aesthetically attractive objects and products. Second, this generates an incentive system in which there is a disproportionate payoff in staying a 'local minimum' where the projects are very simple but at the same time very admired by external observers. Settling for simple projects is a temptation that educators have to avoid at all cost. The non-triviality of navigating these new incentive systems was one of the important lessons learned in these early workshops.

4.2. The Upside-down Roller Coaster, or the Power of Despair

Before coming to one of the digital fabrication workshops, John, Tyler, and Bob found themselves brainstorming about what to build. One of their oddest ideas was to build a roller coaster in their backyard. After the first few days of the workshop they decided to tackle it – but clearly more as a playful thing do to, without much hope of actually building it. Their first step was to scale down the project from a backyard to a 'tabletop' rollercoaster. They then imagined that the process would be quite straightforward: designing the tracks on a vector-drawing software, 'printing' using the laser cutter, and assembling everything.

When they started the design, the first problem came about: how to make curved tracks with uniform width? They realized that they could not just use any type of lines to curve the track, as an uneven track would cause the car to get stuck. Their first challenge was to solve a geometrical puzzle: Should they make tracks using the freehand tool? Bezier curves? Other kinds of curves? Should they create two perfect arcs? Should the arcs have the same radii?

In such an environment, there are no right answers, so these debates take a long time. After a day of discussions and experimentation, they ended up using arcs to create the smoothest possible curve while retaining the width of the track, and printed them out on the laser cutter. However, they realized that they had another problem in hands: the sharpness of the turns, which would make the car 'lose most of its speed'. After much bending and warping of the track, they eventually decided to scrap this design and start from scratch. Tyler worked on the new design, now with much wider turns, which seemed to work better (see Figure 1).

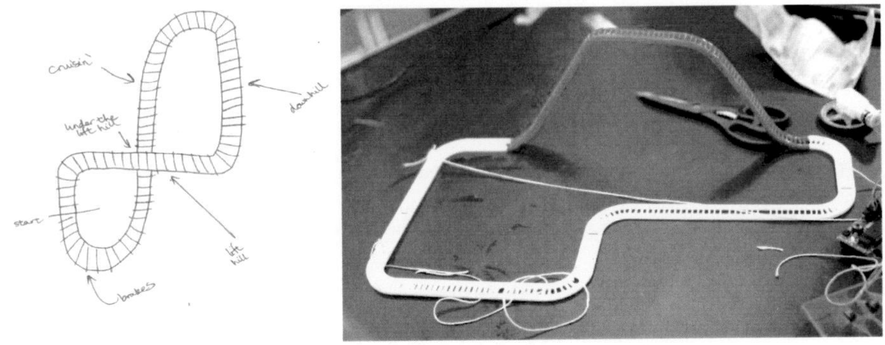

Figure 1: The initial plan for the rollercoaster track (left), and the second design (right) (Source: Photography by Paulo Blikstein).

Another problem that presented itself was the car. Several ideas were proposed for its design. Initially students were considering a normal rollercoaster car that would just roll on top of the track. They tried several designs, with and without wheels, but soon realized that friction was again making the car lose speed. After much brainstorming, they came up with a new idea: a hanging car under the track, which would have much less contact area and thus minimize friction. The group then designed the hanging car and printed it on the 3D printer. However, another problem arouse: students realized they would have to suspend the track so that the car would not touch the table, and could not think of a reasonable way to accomplish it and maintain balance. They went back to the car-on-top idea, now with a better design for the wheels. After a day of hard work, they thought they had the final design, and printed everything. The first test was a disaster: the car did not have enough power to make it around the entire track. The original idea was to rig a device to bring the car to the top of the big hill, and let gravity do the rest. Students attempted several methods to bring the car to the top of the hill, from using magnets to attempt to pull the car up the track, to using a string towrope to pull it up the hill. After hours of failed experiments, they came to an even more devastating realization. Gravity would not provide enough energy to make the car go around the entire track – too much momentum was lost in turns and due to friction, and the plastic car was too light to accumulate enough potential energy. They gave up on gravity, and started considering other options. Their next idea was to rig a sail up on the car and use a battery-powered propeller to provide wind. But they again ran into problems, first with attaching the sail, then with the issue that the force applied to the sail pushed the nose of the car down without moving it. They realized this was due to one of two elements: either the fact that they were using a central wheel system, thus not providing stability in the front or back, or that the wheel system was not providing enough grip, thus the car was not stable enough. Students threw the design out the window, and one team member gave up on the project.

At this point, the rest of the team was desperate, asking everyone in the lab for ideas and help. They had spent almost two weeks on the project and nothing seemed to work – frustration was in the air. Some facilitators volunteered to help and come up with new ideas, and when just about everything seemed to be a failure, one revolutionary idea emerged: instead of having the car go around the track, why not make the track go around the car? The main problems arose from the fact that it was too difficult to power a tiny car without any motors, since the car was too small for that. So what about turning the problem on its head, and move the track instead? The suggested a design would treat the track as a puppet, with strings attached to each of its corners, and motors pulling them up and down – then they could easily use gravity again. Students took on the challenge, printed another track, created an acrylic frame, attached motors and programmed the GoGo Board (Sipitakiat et al. 2004), and in a few hours had a working prototype.

This episode illustrates the working dynamics in a digital fabrication lab or maker space along many dimensions. First, it promoted contextualized encounter with scientific knowledge and lexicon. During the two weeks of the project, students struggled with several physics problems, some of which they knew about but had never seen in real life. Their dialogue, which initially was about the 'losing speed', became increasingly complex, rigorous, and compliant with the lexicon of physics: for example, 'speed' became 'momentum', and the generic statement that the car was 'losing speed' was later decoupled into 'friction', 'number of turns', 'angle of turns', and 'lack of initial potential energy'. They also identified several causes for friction, and discussed ways to minimize it by reducing the surface area, the friction coefficient, or making the car hang from the track. Physics, engineering, and problem solving were part of the same process, just like when engineers work.

Second, the space was flexible with students' contrasting attitudes towards failure. The narrative of this episode is as baffling and meandering as the project development was. There were no easy answers for the problems that students were facing. The group went through a complex process, filled with frustration, failure, but also exhilarating success in the end. While Tyler, Bob, and John worked together for almost the entire program, they had very different styles in going about their projects. As a team, Tyler's optimism in the face of adversity worked as a great balance to John's aptitude for ideating. While John often drove the start of projects, it was Tyler who would use the inevitable failures to advance their goal. Tyler would often tell John "Things never work the first time, and that's okay". Almost every day they hit a fundamental problem with their design, and consistently came up with means to work through it. While Tyler took the constant setbacks in stride, accepting them as part of the engineering process, John considered them instead as embarrassing failures. Despite these differences, the team showed remarkable perseverance throughout in the project, and was able to use their different approaches to failure as a feature of their collective strategy of problem solving, rather than a difficulty. These students were able to experience realistic engineering design because they had the space and time to fail and try again, and a strong motivation to pursue their own idea. In short-term projects, scripted construction challenges, or time-constrained competitions, the class dynamics would have been radically different, and students would never have been able to experience these dramatic levels of failure and reward. Ultimately, their deep sense of achievement was a consequence of their visceral involvement in the construction of the rollercoaster, and the originality of their design was only possible because of the technical and emotional support that they had in order to withstand extended frustration, shake it up, and go back to the drawing board.

4.3 "The Most Math I Have Ever Learned in a History Class", or the Power of Interdisciplinary Projects

Digital fabrication is typically associated with the learning and practice of STEM disciplines. Laser cutters and 3D printer are 'hard sciences' territory, and supposedly math and science teachers should be the ones primarily involved. In one of the projects in the Lincoln school, however, we had an unlikely scenario: Heather, a history teacher with many years of experience, wanted to bring her four 8th grade classes to the lab. She was not a typical early adopter of a digital prototyping lab – in one of our surveys, she rated herself at the bottom of the scale in knowledge about robotics, mechanical engineering, and computer programming. But Heather was not concerned in training future STEM workers. Her main goal fell within the disciplinary boundaries of History: she wanted her students to learn about great female characters in American history by building historical monuments for them, using 3D printing and laser cutting.

Heather's project illustrates two aspects of the implementation of digital fabrication in classrooms. First, I show how she prepared herself for, and structured the activity. Second, I narrate how a complex and productive 'division of labor' emerged from the project as a result of the interactions between the technical lab teacher and Heather.

Heather had gone through the digital fabrication training workshop and had basic understanding of how most of the equipment worked. She was not comfortable using the laser cutter by herself, though. However, we had set up the lab with a full time technical lab person that could help teachers and students to operate and learn the machines. Thus Heather did not feel she had to master all the tools before starting to work with her students, or that she would be alone with the students exploring all of the unfamiliar territory of the lab.

Even though Heather was not well versed in programming or engineering, she was comfortable using unfamiliar technologies in her classroom. Part of her method was to experiment with the technology 'as a learner' before even starting working with students. Therefore, two days before the start of the digital fabrication activity, she created her own historical monument using the digital fabrication equipment. She went through the entire process herself and understood the challenges and difficulties in building it. Thus when she started working with students, she not only felt more comfortable facilitating the activity, but could also predict bottlenecks and difficulties. She became aware of how much technical expertise was needed to facilitate the project, and realized that the help of the lab's technical coordinator, David, would be crucial. As the activity unfolded, Heather's role evolved to be a project manager, and David's to be a design helper and an equipment operator. Heather set class goals, checked in on the groups, volunteered to laser cut pieces for them in her spare time, and kept track of time. David would sit by the laser cutter most of the time, not interfering with the girls' designs, and acting mostly as a facilitator and consultant. His help was instrumental in moving many groups forward. When the students had a very difficult technical challenge to solve, David had the ability to envision how the entire system should work and give life-saving suggestions. Most students did not have the ability to look at their work as a system, but on interactions between one or two parts at most. David would guide them throughout the process, not taking over or undermining their ideas, but co-designing. The work dynamics that Heather and David put in place was different from a traditional classroom, of course, but it was also a departure from many technology-based after school programs where there is no space for a person with Heather's profile.

Figure 2: One in-progress project with the grid clearly visible (left), and some of the projects done by students (middle, right) (Source: Photography by Paulo Blikstein).

In a robotics workshop, each group has its own equipment and work autonomously. In a digital fabrication environment, however, the work is centralized in just one or two machines. The question, then, is deciding on the side of efficiency (one specialist operating the machine for everyone), or equity (everyone operating the machine). In Heather and David's model, the workflow was faster, but there was doubt if the girls were actually learning while watching David work through problems, or if they were simply relieved to have something done for them, and if they could come up with elegant solutions without him.

This assembly-line division of labor made it possible for students to get their parts cut in 50-minute periods. However, there were unintended consequences to this scheme. Some students may have prematurely aborted design elements that they deemed too difficult to do on their own, given the time constraint. In addition, the amount of experience Heather's students had with fabrication varied from group to group. It seemed like there was a place for every student on the spectrum to fit in. Some groups required more technical (and mathematical) help than others, but the students all seemed to be in charge of the creative part of their designs. However, it was unclear whether David was being helpful by doing some of the more complex calculations for the students or simply passing the information along to the students. At this point the aim was efficiency; and it could have undermined students' willingness to persist through difficult problems.

Heather also made changes to her own activity design. Instead of a completely open-ended project, she introduced some structure: the wooden base of the historical monuments would be standardized (a 15 · 15 square grid with a 3/16":1 ft scale, see Figure 2, left).

Heather's activity design, which was intended to give students a safe starting point for their projects, had an unexpected consequence as well. The activity, which was originally a history project, suddenly became a sophisticated mathematics project as well. When Heather standardized the base and assigned strict dimensions to it, she foreground one aspect of the activity that would have been overlooked by many students: measurement. All of a sudden, the objects had to fit the base and their relative sizes had to be exact. Students did not want to have a park bench be the same size as a person, and they knew they could not tinker with the dimensions of objects after the fact. In the post-interviews, students were positively shocked with how much math they had to learn to accomplish the history project.

The History Monuments Project unwittingly illustrated some additional principles of digital fabrication in schools. First, digital fabrication introduces a new type of 'workflow'. Different from a science lab or a robotics workshop, in which each group works autonomously with one kit, in a FabLab there is just one laser cutter. This generates pressure for productivity and division of labor that could be either productive or disempowering. In Heather's case, the scheme was mostly productive in which it enabled students to focus on the creative part of the project, not having to deal with the specifics of the software and the laser cutter. In fact, the division of labor was also a crucial enabler for the project to happen within the four 50-minute time slots that she had. However, we also observed that this scheme could easily turn into a disempowering arrangement when students realize that they are too dependent on the facilitators and cannot create the more complex designs by themselves – all the hard work is done by eager facilitators racing against the school bell. Second, the environment was conducive to unlikely interdisciplinary projects: The making of a physical project will always entail some engineering work. Despite the fact that students were working on a history-themed project, they ended up having to explore multiple topics in mathematics such as measurement, scale, and proportion, both in two and three dimensions. In the same way that the rollercoaster students encountered physics in authentic ways, the monuments students were seeing mathematics everywhere.

4.4 The Robotic Flute Player, or the Demise of Constant Airflow

Max, a high school student in Moscow, was not an engineering type. He was passionate about music – Bach in particular. In one of our first meetings, he told the facilitators his childhood dream was to build a robot that could play Bach – thus his interest in digital fabrication, although he had no idea where to start. After a week, he had learned how to laser cut, program, control motors and sensors, and had an incredible prototype of the flute. It was not yet good enough for him. The workshop was over, but now finishing the flute was his personal project, so he kept coming to the lab for two months, several times a week. In the end, he built a flute with 12 servomotors, a highly complex control mechanism, and was able to play some simple Bach melodies by programming the microcontroller board. He took this project to the National Science Exhibition, a very competitive event in Russia with hundreds of students from all over the country, and won 3rd place. It seems like a success, but the competition was not the most important part of the story.

Apart from controlling the robotic 'fingers', Max also wanted a machine to blow air into the flute. For several weeks, he tried many solutions – foleys, pumps, vacuum cleaners, and even complex piping systems. After countless experiments and redesigns, he finally found a way to blow a consistent amount of air at just the right angle to produce sound. He decided to use an inverted vacuum cleaner with a series of polymer-cast pipes, which he made himself. Max turned the system on, started the servo motor system, and waited for the contraption to play Bach – when something remarkable happened. Even though the system was working as planned, it did not sound like a Bach piece. Something was off: the movements were correct, the air was flowing steadily, but it was not what he expected. After much reflection with the facilitators, Max finally understood the problem. No human flute player would have a steady flow of air – regulating the airflow is exactly the craft of musicians, who interpret the melody in their own way, emphasizing and highlighting different parts. Bach pieces sound weird when played by a robot because there is no interpretation, just an automatic execution with constant airflow.

Max was disappointed but also extremely happy, too – by building a robotic flute, he had learned a lot about engineering, but the main lesson was about music interpretation, and the true craft of a musician. His episode illustrates, again, the integrated nature of projects in the lab, where there is no real boundary between disciplines. But even more importantly, it shows a crucial component of the lab's success: attracting students who would not traditionally see themselves as engineers or scientists. Since the lab was architected (and advertised) as a place for invention – and not for 'building robots' or making lights blink – even students like Max felt compelled to try something. His contribution, as a musician/engineer, made the environment more diverse and intellectually rich, attracted even more students, and infused unexpected ideas into other students' work.

5. Digital Fabrication and Making: The Ultimate Construction Kit

In this chapter, I first told the story of the rise and fall then rise again of the making and building in education, discussed the theoretical underpinnings of project-based, student-centered learning, and presented the work of John Dewey, Seymour Papert, and Paulo Freire. I discussed how digital fabrication brings unique tools to progressive educators, and presented four prototypical episodes that exemplify the advantages and perils of FabLabs in schools.

These examples highlight five important design principles:

(a) The 'Keychain Syndrome': since digital fabrication machines might generate aesthetically-pleasing pieces with little effort, educators should shy away from quick demonstration projects and push students towards more complex endeavors;

(b) The power of despair and visceral involvement: FabLabs provide an environment for unprecedented visceral design experiences, multiple cycles of design, and new levels for both frustration and excitement, which students normally do not undergo in their normal school experience;

(c) Powerful interdisciplinary projects: the artificial boundaries between disciplines are completely reconfigured in the lab. History and mathematics become closely related, and so do music and robotics, and this richness results in a more diverse and accepting intellectual environment;

(d) Contextualized learning in STEM: students have the opportunity to come across several concepts in engineering and science in a highly meaningful, engaging, and contextualized fashion. Abstract ideas such as friction and momentum become meaningful and concrete when they are needed to accomplish a task within a project; math becomes a necessity in a history project.

(e) Intellectualization and re-evaluation of familiar practices, rather than the replacement of existing ones (Blikstein 2008): Students bring their own familiar practices to the lab (craft, construction, carpentry), and those practices get augmented using socially-valued tools such as computation and mathematics. The malleability of the equipment and the pedagogical space in the lab makes the augmentation and embracement of such practices feasible, generating an environment that values multiple ways of working.

Despite the potential of digital fabrication labs and 'making' in education, educators and scholars must remember that, as Seymour Papert would say, the real power of any technology is not in the technique itself or in the allure it generates, but in the new ways of personal expression it enables, the new forms of human interaction it facilitates, and the powerful ideas it makes accessible to children.

Acknowledgement

Special thanks to all the students and their incredible work. This was a truly collaborative work of many people: the Stanford team, Marcelo Worsley, Shima Salehi, Engin Bumbacher, Claire Rosenbaum, Kathryn Papadopoulos; the Castilleja School team, Angi Chau, Diego Fonstad, and Heather Pang; the MPEI 1502 school team (Moscow), Irina Krasnova, Anton Vasiliev, and Nadezhda Alekseeva; the East Palo Alto Academy High School team, Derek Ang, Guy Mathews; the Darunsikkhalai School for Innovative Learning team (Bangkok), Nalin Tutiyaphuengprasert, Arnan Sipitakiat, and all the students and their incredible work. This work was funded by the NSF CAREER award #1055130, Schlumberger Excellence in Educational Development Foundation, Euan Baird, the Levin Fund, the Lemann Foundation, and the Suksapattana Foundation.

REFERENCES

Abrahamson, D, Blikstein, P, & Wilensky, U 2007, 'Classroom model, model classroom: Computer-supported methodology for investigating collaborative-learning pedagogy', in Chinn, C, Erkens, G & Puntambekar, S (eds.), *Proceedings of CSCL Computer Supported Collaborative Learning Conference*, vol. 8, no. 1, pp. 46-55.

Astrachan, O, Hambrusch, S, Peckham, J & Settle, A 2009, 'The present and future of computational thinking', *Paper presented at the 40th ACM technical symposium on computer science education*, Chattanooga, TN, USA.

Atkin, J 1990, 'Teach science for science's sake: for global competitiveness, try technology', *Education Week*, vol. 10, no. 32.

Berland, M 2008, 'VBOT: Motivating Computational and Complex Systems Fluencies with Constructionist Virtual/Physical Robotics', PhD thesis, Northwestern University, Evanston, IL.

Berland, M & Wilensky, U 2006, 'Constructionist collaborative engineering: Results from an implementation of PVBOT', *Paper presented at the Annual meeting of the American Educational Research Association*, San Francisco, CA, USA.

Blikstein, P 2008, 'Travels in Troy with Freire: Technology as an Agent for Emancipation', in Noguera, P & Torres CA (eds.), *Social Justice Education for Teachers: Paulo Freire and the possible dream*, Sense, Rotterdam, Netherlands, pp. 205-244.

Blikstein, P 2010, 'Connecting the science classroom and tangible interfaces: the bifocal modeling framework', *Proceedings of the 9th International Conference of the Learning Sciences – Volume 2*, Chicago, IL, USA.

Blikstein, P, Fuhrmann, T, Greene, D & Salehi, S 2012, 'Bifocal modeling: mixing real and virtual labs for advanced science learning', *Proceedings of the 11th International Conference on Interaction Design and Children*, ACM, Bremen, Germany.

Buechley, L 2006, 'A Construction Kit for Electronic Textiles', *Proceedings of the IEEE International Symposium on Wearable Computers*, Montreux, Switzerland.

Buechley, L & Eisenberg, M 2008, 'The LilyPad Arduino: toward wearable engineering for everyone', *IEEE Pervasive Computing*, 7, pp. 12-15.

Buechley, L, Eisenberg, M, Catchen, J & Crockett, A 2008, 'The LilyPad Arduino: using computational textiles to investigate engagement, aesthetics, and diversity in computer science education', *ACM*, pp. 423-432.

Cavallo, D 2000, 'Emergent Design and learning environments: Building on indigenous knowledge', *IBM System Journal*, 39 (3&4), pp. 768-781.

Cooper, S, Dann, W & Pausch, R 2000, 'Alice: A 3-D Tool For Introductory Programming Concepts', *Proceedings of 5th Annual CCSC Northeastern Conference*, Ramapo, NJ, April 28-29.

Dewey, J 1902, *The Child and Curriculum*, University of Chicago Press, Chicago, IL.

diSessa, AA 2000, *Changing minds: computers, learning, and literacy*, MIT Press, Cambridge, MA.

Dym, CL 1999, 'Learning Engineering: Design, Languages, and Experiences', *Journal of Engineering Education*, pp. 145-148.

Edelson, D 2000, *My World GIS*, PASCO Scientific, Evanston, IL.

Feisel, LD & Rosa, AJ 2005, 'The Role of the Laboratory in Undergraduate Engineering Education', *Journal of Engineering Education*, 94 (1), pp. 121-130.

Freire, P 1974, *Pedagogy of the oppressed*, Seabury Press, New York.

Freudenthal, H 1973, *Mathematics as an educational task*, Reidel, Dordrecht.

Fröbel, F & Hailmann, WN 1901, *The education of man*, D. Appleton, New York.

Gershenfeld, N 2007, *Fab: the coming revolution on your desktop - from personal computers to personal fabrication*, Basic Books, New York.

Grinter, LE 1955, *Report on Evaluation of Engineering Education*, ASEE, Washington, DC.

Illich, I 1970, *Deschooling society*, Harper & Row, New York.

Levy, F & Murnane, RJ 2004, *The new division of labor: How computers are creating the next job market*, University Press, Princeton.

Martin, F 1994, 'Circuits to Control: Learning Engineering by Designing LEGO Robots', PhD thesis, MIT, Cambridge, MA.

Martin, F & Resnick, M 1993, 'Lego/Logo and electronic bricks: Creating a scienceland for children', in Ferguson, DL (ed.), *Advanced educational technologies for mathematics and science*, Springer-Verlag, Berlin, Heidelberg.

Mikhak, B, Lyon, C, Gorton, T, Gershenfeld, N, McEnnis, C & Taylor, J 2002, *Fab Lab: An Alternative Model of ICT for Development, "development by design" (dyd02)*, ThinkCycle, Bangalore.

Millner, A & Resnick, M 2005, 'Tools for Creating Custom Physical Computer Interfaces', *Proceedings of the 7th international conference on interaction design and children*, Boulder, CO.

Montessori, M 1964, *The advanced Montessori method*, Bentley R., Cambridge MA.

Montessori, M 1965, *Spontaneous activity in education*, Schocken Books, New York.

National Research Council 1999, *Being Fluent with Information Technology*, The National Academies Press.

National Research Council 2002, *Technically Speaking: Why All Americans Need to Know More About Technology*, The National Academies Press.

Papert, S 1980, *Mindstorms: Children, computers, and powerful ideas*, Basic Books, New York.

Papert, S 1987, 'Computer Criticism vs. Technocentric Thinking', *Educational Researcher*, vol. 16, no. 1.

Papert, S 1991, 'Situating Constructionism', in Papert, S & Harel, I (eds.), *Constructionism*, MIT Press, Cambridge, MA.

Perner-Wilson, H, Buechley, L & Satomi, M 2011, 'Handcrafting textile interfaces from a kit-of-no-parts', *Proceedings of the fifth international conference on Tangible, embedded, and embodied interaction*, ACM, pp. 61-68.

Raffle, HS, Parkes, AJ & Ishii, H 2004, 'Topobo: a constructive assembly system with kinetic memory', *Proceedings of the SIGCHI conference on Human factors in computing systems*, ACM, pp. 647-654.

Resnick, M, Maloney, J, Monroy-Hernández, A, Rusk, N, Eastmond, E, Brennan, K, Millner, A, Rosenbaum, E, Silver, J, Silverman, B & Kafai, Y 2009, 'Scratch: programming for all', *Commun*, ACM 52, 11/2009, pp. 60-67.

Rousseau, JJ 1961, *Emile*, Dutton, New York.

Schweikardt, E & Gross, MD 2006, 'roBlocks: a robotic construction kit for mathematics and science education', *Proceedings of the 8th international conference on Multimodal interfaces*, ACM, Banff, Alberta, Canada.

Sheppard, S & Jenison, R 1997, 'Examples of Freshman Design Education', *International Journal of Engineering Education*, vol. 13, no. 4, pp. 248-261.

Sipitakiat, A 2000, 'Digital Technology for Conviviality: making the most of learners' energy and imagination', MSc thesis, MIT, Cambridge, MA.

Sipitakiat, A, Blikstein, P & Cavallo, DP 2002, 'The GoGo Board: Moving towards highly available computational tools in learning environments', paper presented at the Interactive Computer Aided Learning International Workshop, Villach, Austria.

Sipitakiat, A, Blikstein, P & Cavallo, DP 2004, 'GoGo Board: Augmenting Programmable Bricks for Economically Challenged Audiences', *Proceedings of the International Conference of the Learning Sciences*, Los Angeles, USA, pp. 481-488.

Swetz, F & Smith, D 1987, *Capitalism and arithmetic: The new math of the 15th century*, Open Court Publishing, La Salle, Illinois.

Sylvan, E 2005, 'Integrating Aesthetic, Engineering, and Scientific Understanding in a Hands-on Design Activity', *Proceedings of the Interaction Design for Children*, ACM, Boulder, CO.

Tryggvason, G & Apelian, D 2006, 'Re-engineering engineering education for the challenges of the 21 st century', *JOM*, vol. 58, no. 10, pp. 14-17.

von Glasersfeld, E 1984, 'An Introduction to Radical Constructivism', in Watzlawick, P (ed.), *The Invented Reality*, Norton, New York.

Wilensky, U 1999, *NetLogo. Evanston, IL: Center for Connected Learning and Computer-Based Modeling*, viewed 15 February 2013 <http://ccl.northwestern.edu/netlogo>.

Wilensky, U & Papert, S 2010, 'Restructurations: Reformulating knowledge disciplines through new representational forms', in Clayson, JE & Kalas, I (eds.), *Constructionism 2010*, Paris, France.

Wilensky, U & Stroup, W 1999, 'Learning Through Participatory Simulations: Network-Based Design for Systems Learning in Classrooms', paper presented at the Computer Supported Collaborative Learning Conference, Stanford University, California.

Yasar, O & Landau, RH 2003, 'Elements of computational science and engineering education', *SIAM review*, vol. 45, no. 4, pp. 787-805.

URBAN DEVELOPMENT WITH FABLABS

AXEL SYLVESTER, TANJA DÖRING

Figure 1: The picture shows the entrance of the FabLab in St. Pauli. The place where the FabLab is located is a meeting point in the neighborhood, where for example flea markets take place on Saturdays (Source: Photography by Tanja Döring).

1. Introduction

Our society is constantly changing and so are our cities. Urban development addresses these changes and analyzes aspects of land and building usage as well as the role their professional work plays in the lives of habitants. In cities, living and working are often converged (e.g., when working self-employed). Due to technical developments, this also concerns the production of goods nowadays. These technical developments leverage the way for a serious counter model to centralized mass production: the scattered digital fabrication. FabLabs are among those kinds of facilities where this development is currently emerging. But how can FabLabs be implemented and sustained in an urban context so that they are not only separate facilities for single users? The answer lies in embedding the labs into the local community on the one hand and into the larger network of labs beyond the local community on the other hand. In this article, we first describe the overall potential we see for the role of FabLabs in a larger network of innovation and, second, we apply urban development patterns to FabLabs on the basis of experiences from the FabLab Fabulous St. Pauli in Hamburg (Germany).

2. FabLabs as Hubs in Local Embedded Economies and Center-periphery Networks

In economics, there are two main opposing models on how to organize production, the centralized mass production (in one big production facility) and the locally organized, decentralized scattered production in small production facilities. The first evolved during the Second Industrial Revolution out of the rationalization aspects, while the latter is a counter model that gained more practical relevance in the last years, as new production techniques allowed to produce highly specialized high-tech components in small production facilities, in small scales and often with less investment needed (this development is sometimes referred to as the "third industrial revolution" (The Economist 2012).

Both models have a number of advantages and disadvantages. The main argument for centralized mass production is that, in case the demand for a product is big enough and there is no need in customizing a product, the production on a large scale is often more efficient. The processes on a large-scale production are optimized and thus cheaper. As the players are known and the number of production facilities is not too big, the control of industry standards by official organizations is – at least theoretically – possible. Furthermore, in a big production facility, risks, pollution and waste appear in one place. However, centralized mass production also has a number of large disadvantages. Among these are the long transportation ways and high logistic efforts involved that are only affordable as long as energy costs do not explode (while an explosion of energy costs is, for example, one of the future scenarios in the theories about the Third Industrial Revolution). Other downsides of mass production are the big investments needed, the large risks such investments pose, and a non-transparent production with often inadequate conditions for employees. Schumacher called this technology of mass production "inherently violent, ecologically damaging, self-defeating in terms of non-renewable resources and stultifying for the human person" (Schumacher 1973, pp. 126-127) and argued for a decentralized production by the masses.

Compared to mass production, decentralized production has many advantages. First, the needed investments and the risks involved are smaller, and the distribution of products happens regionally so that transportation costs are comparably low. Second, it results in a high product diversity, local needs can be addressed easily, and the overall production is more robust against disruptions. Third, production processes are likely to be more transparent and more ethical (i.e, no exploitation of labor), as well as the ingredients and the materials a product is made of. Moreover, customers usually trust and appreciate local products more, especially if they know the producers. For a society, a decentralized production also provides means for more democratic production processes and technologies.

For a sustainable urban development, the support of local embedded economies that realize decentralized production is essential, as an export-oriented economic promotion does not solve local problems. While the latter rather reinforces urban displacement, local economies have the potential to prevent and overcome the isolation of underprivileged urban quarters by empowering their citizens and connecting them to actors in other areas of the city and beyond. (Läpple 2005) These connections are part of larger networks which form an important basis for innovation.

According to Katzmair (2008), who analyzed the Austrian network of centers for technology, ideal networks follow a "center-periphery model" (Katzmair 2008; own translation). This model is a combination of the "single peaked network" (Katzmair 2008; own translation) that relies on one strong actor in the center, which attracts further actors, and the "multi peaked network" (Katzmair 2008; own translation) that consists of many connected, equally strong actors. In an ideal "center-periphery network" (Katzmair 2008; own translation), a strong central actor is highly connected to many heterogeneous actors in the periphery, who are well connected with each other, too. A high degree of diversity is a positive factor for innovation. While this innovation mainly happens at the periphery, the center has the capability to transfer it from the periphery to the market. Referring to Schumpeter (1912), Katzmair argues that these centers are important 'hubs' that stabilize an innovation landscape, "make networks more efficient and manageable and offer especially the periphery a gradual orientation towards good practices in the particular area (successful actors apparently serve as role models for other actors, they imitate them, learn from them – an important factor for the regional diffusion of good practices)" (Katzmair 2008; own translation).

We argue that FabLabs have a strong potential to serve as hubs in these networks of innovation (following the center-periphery model), and to support local embedded economies because of the following reasons:

- FabLabs are globally connected and also serve as a local hub at the same time, where actors meet, connect, and exchange knowledge and ideas.
- FabLabs provide access to tools for digital fabrication beyond production for personal needs. Fabrication in FabLabs has a professional character.

- The actors involved in FabLabs are committed to an Open Source and Open Design mindset and are driven by the awareness of global economical and ecological problems.

(1.) From the beginning of the FabLab movement, the network between FabLabs has been at the heart of the whole idea. The exchange of experiences about how to set up and run FabLabs, the further development of Open Source machines and software, the co-developing of ideas and products – all of this would not be possible without the network between FabLabs (and other communities, of course), which already spans around broad parts of the world and is constantly growing. Thus, besides its status as a workspace, the local FabLab also serves as a meeting point and local network, as well as a gateway to a strong worldwide network. In this network, FabLabs have the potential to provide centers for innovation. They can provide hubs for the numerous and diverse innovative actors in the periphery and provide platforms and the network to distribute innovative ideas and products. Especially as the density of FabLabs will increase, FabLabs will also form accessible hubs in suburbs and rural communities. Furthermore, the educational aspects of FabLabs, which have been addressed by the Fab Academy, focus on the network. Gershenfeld (2009) describes it as "a network rather than a place, with teachers and students in FabLabs around the world linked by broadband video, shared online information, and common technical capabilities" (Gershenfeld 2009). Thus, we see the center-periphery interface implemented as a central idea of FabLabs that always link the local community to the regional and the global community.

(2.) Besides being an educational and innovational hub, FabLabs also are a facility for peer production. We have seen many cases where the breakaway of big industries and the loss of jobs led to problems for an entire region (e.g., no income, no perspective for development, abandoned buildings, unmaintained infrastructure). City officials often see the resettlement of new big industry as the only solution. The sharing of machines, tools and infrastructure (e.g., in a FabLab) by small players is a considerable alternative. This is a different business model, which requires more organization, but also should be more resilient and does not require much investment in infrastructure. One of the advocates of this idea is the philosopher Frithjof Bergmann, who experienced the crisis in the automotive industries in Detroit in the 1970s and 1980s and the problems that occurred for the region. He started to promote a new way of work that included personal fabrication for the own community as an efficient and considerably low-priced and thus competitive alternative and founded the first Center of New Work in Flint (Michigan) in 1984 (Bergmann 2004). FabLabs provide the infrastructure for personal fabrication and offer a broad range of machines to process various materials. People, who might have been active in traditional crafts enterprises before, meet and share the machines, and collaboratively fabricate products. This infrastructure is beyond personal hobby handicraft work: fabrication in FabLabs has a professional character, which can be decisive in encouraging production for real markets as well.

(3.) "Think globally, fabricate locally" (Gershenfeld 2012) is a central credo in the FabLab community (as well as in other maker and hacker communities). This credo implies global thinking on many levels: sustainability and environmental awareness, preventing discrimination and unequal opportunities, spreading knowledge and skills, co-development and distribution of hardware and software products and machines etc.. Regarding all these aspects, a strong focus on innovation can be found in the FabLab community. Among the most important driving forces of innovation are Open Source development and Open Design. It would simply not be possible to develop machines and software or to design products to the given extent and at the current rate without the Open Source network. Nowadays, Open Source is favored by many, as Open Source machines, products, and software can be extended and branched, controlled or personalized easily by a community or single persons. Open Source allows for a fast and agile development since many people can contribute, and it relies on support as well as on the community (give-and-take). Patents would hinder this process; additionally, they would be complicated and expensive for single persons and thus often are not feasible anyway. While the knowledge how to build high-tech machines, software, and other products used to be kept secret among involved employees of a limited number of companies, this knowledge is now decentralized and available for anyone interested. Moreover, due to the advances hardware and software undergoes, the fabrication process is becoming easier. Overall, using a FabLab is different compared to using traditional service providers: the underlying model is not to let things be developed and produced somewhere; it is even beyond DIY – it is DIWO: "Do It With Others" (McCue 2012).

In the following, we discuss sample aspects of FabLabs as hubs in local embedded economies along urban development patterns, and provide experiences from our FabLab Fabulous St. Pauli in Hamburg, Germany. The FabLab was founded in 2011 and is located in a community center. One of the two open days per week is an open lab day where anyone can use the lab for free, only covering expenses for materials.

3. The Application of Urban Development Patterns in FabLabs: Experiences from the 'Fabulous St. Pauli'

In the 1970s, the architect Alexander and his colleagues wrote a series of three books in which they describe "detailed patterns for towns and neighborhoods, houses, gardens, and rooms" (Alexander et al. 1977, p. ix). In the following, we introduce six patterns that we regard as closely related to the way FabLabs should be implemented in a city, i.e., "mosaic of subcultures", "community of 7000", "scattered work", "work community", "network of learning" and "shopfront school" (Alexander et al. 1977), and will relate them to activities in our FabLab.

The pattern mosaic of subcultures addresses the problem of "the homogeneous and undifferentiated character of modern cities [that] kills all variety of life styles" (Alexander et al. 1977, p. 43), and suggests the establishment of a large number of relatively small subcultures: "People can choose the kind of subculture they wish to live in, and can still experience many ways of life different from their own. Since each environment fosters mutual support and a strong sense of shared values, individuals can grow" (Alexander et al. 1977, p. 43). In order to support the development of subcultures Alexander et al. suggest: "Do everything possible to enrich the cultures and subcultures of the city, by breaking the city, as far as possible, into a vast mosaic of small and different subcultures, each with its own spacial territory, and each with the power to create its own distinct lifestyle. Make sure that the subcultures are small enough, so that each person has access to the full variety of life styles in the subculture near his own" (Alexander et al. 1977, p. 50).

The FabLab in Hamburg St. Pauli is located in a very dense and diverse old area of the city. Until the end of the 19th century, the neighborhood was located outside of the city walls. This led to a situation where a very diverse and unadapted neighborhood came to existence and this characteristic can still be felt nowadays. A specific of the neighborhood is that areas for entertainment, bars, recreation, production, urban gardening, media agencies, culture, spirituality, sport, and living are located very close to each other. This diversity has led to a fine grain pattern of subcultures within the area. The people who initiated and run the FabLab in St. Pauli are very familiar with these structures because they live and work in this area. As we were aware of these subtle subcultural circumstances, we knew that a proper location for a FabLab in the neighborhood was important. Before we had a lab space, we promoted the concept by organizing hands-on workshops in various scenes – in a mobile FabLab on the street in cooperation with a citizens' initiative, at an adventure playground, at technical clubs (Linux Café, Chaos Computer Club), at a sustainability trade fair, in a cultural center etc.. What we achieved by this was that we were not postmarked to be belonging to a specific subculture (e.g., artists, people with mainly commercial interest, political activists) within the city. The FabLab in Hamburg is now temporarily located in a community center called Centro Sociale. Visitors of the lab are quite diverse, ranging from inventors to designers, from technicians to newbies, from experienced engineers to students. Poor and rich people come to the lab, the age range is from kids to retired people. Male and female are in numbers more or less equal.

The Centro Sociale was initiated and is run by a group that evolved directly out of the nearby neighborhood, because there was a strong need for a public community center and meeting place for various groups, and the city officials did not meet these requirements (one can see the community center from outside in Figure 1). This situation is described by another pattern, the community of 7000. The authors state that "individuals have no effective voice in any community of more than 5000-10,000 persons [and give advice to] decentralize city governments in a way that gives local control to communities of 5,000 to 10,000 persons. [...] Give each community the power to initiate, decide, and execute the affairs that concern it closely: land use, housing, maintenance, streets, parks, police, schooling, welfare, neighborhood services" (Alexander et al. 1977, pp. 73-74).

A further problem that should be addressed is the way we work. With the pattern scattered work Alexander et al. address "the artificial separation of houses and work" (Alexander et al. 1977, p. 52). They argue that "concentration and segregation of work leads to dead neighborhoods" (Alexander et al. 1977, p. 52), and advise that workplaces shall be within walking distance of children and families. In another pattern, Alexander et al. suggest to build work communities: "If you spend eight hours of your day at work, and eight hours home, there is no reason why your workplace should be any less of a community than your home" (Alexander et al. 1977, p. 223). They propose to design each one as "a collection of smaller clusters of workplaces which have their own courtyards, gather around a larger common square or common courtyard which contains shops and lunch counters. The total work community should have no more than 10 or 20 workplaces in it" (Alexander et al. 1977, p. 225). This exactly describes the situation for the current location of the FabLab in Hamburg.

Over the course of the last years, a move towards a more modular, project-based, and self-employed work, e.g., in the Internet sector, has taken place. Beyond working in the domain of Internet and New Media, self-employed people increasingly also work on the creation of physical artifacts and hardware devices nowadays. While there are shared office spaces (co-working spaces) for office workers, workshops for the creation of physical artifacts are rare, expensive, fairly isolated, and in most cases initiated by individual motivation. Emergence is seldom. We see a FabLab as a facility that addresses such topics. It is a meeting place, for example, as well as a place to share tools, machines, and equipment. The relatively small-scale character of FabLabs and the FabLab as a decentralized production facility support the requirements that are described in the "scattered work" (Alexander et al. 1977, p. 222) pattern and the pattern "work community" (Alexander et al. 1977, p. 222).

Two more patterns are closely related to the way FabLabs should function in our point of view: network of learning and shopfront school. Network of learning deals with the problem that "in a society, which emphasizes teaching, children and students – and adults – become passive and unable to think or act for themselves. Creative, active individuals can only grow up in a society which emphasizes learning instead of teaching" (Alexander et al. 1977, p. 100). To overcome this, the authors demand to "decentralize the process of learning and enrich it through contact with many places and people all over the city" (Alexander et al. 1977, p. 102). Children shall have the opportunity to learn from professionals, from other children, in industrial workshops, from old people etc.. A FabLab is an ideal place for such encounters. The effort made by instructors and practitioners in FabLabs and elsewhere could be compensated by a payment in vouchers (Alexander et al. 1977, p. 102). If educational programs for children are implemented, while other people use the lab for professional purposes, then there is a good chance that kids get to know role models and perspectives for their own professional lives.

4. Conclusion

This article addressed the value of FabLabs in the context of urban development and local economies. Due to their integration into a global network, their provision of a professional environment for digital fabrication and the innovative and Open Source-oriented mindset of their actors, FabLabs provide the means for hubs in center-periphery networks of innovation. For a successful implementation of FabLabs for urban development, it is useful to take a closer look at patterns from architecture and urban planning. We discussed six of these patterns and their application in the context of our FabLab in St. Pauli. Although these patterns provide helpful guidance, we believe that a top-down implementation of FabLabs is not likely to be successful – FabLabs need to grow and evolve with the local people. FabLabs generally attract all kinds of people from different backgrounds rather than one specific target group, but to really achieve a good mix in a FabLab, the activity of a local community is needed. On the whole, we are convinced that FabLabs have a big potential for supporting innovations and innovative ways of developing and producing goods. With FabLabs, a scattered network of strong local innovation networks are emerging that might promote new centers in the long run.

References

Alexander, C et al. 1977, *A pattern language*, Oxford University Press, New York.

Bergmann, F 2004, *Neue Arbeit, Neue Kultur*, Arbor-Verlag, Freiamt im Schwarzwald.

Gershenfeld, N 2009, 'Is MIT obsolete?', in *SEEDMAGAZINE.COM*, viewed 2 November 2012, <http://seedmagazine.com/content/article/is_mit_obsolete/>.

Gershenfeld, N 2012, 'How to make almost anything. The digital fabrication revolution', in *Forreign Affairs*, November/December 2012, viewed 2 November 2012, <http://www.foreignaffairs.com/articles/138154/neil-gershenfeld/how-to-make-almost-anything>.

Katzmair, H 2008, *Die österreichischen Technologiezentren: Beobachtungen und Überlegungen aus dem Blickwinkel der Netzwerkanalyse und Komplexitätsforschung*, viewed 2 November 2012, <http://www.vto.at/media/portals/1/oesterr.Technologiezenren2008.pdf>.

Läpple, D 2005, 'Lokale Ökonomie', in Ritter, EH (ed.), *Handwörterbuch der Raumordnung*, Akademie für Raumforschung und Landesplanung, Hannover.

McCue, TJ 2012, 'Do it with others – maker community manifesto', in *Forbes.com*, viewed 2 November 2012, <http://www.forbes.com/sites/tjmccue/2012/06/12/do-it-with-others-maker-community-manifesto>.

Schumacher, EF 1973, *Small is beautiful: Economics as if people mattered*, Blond & Briggs Ltd., London.

Schumpeter, J 1912, *Theorie der wirtschaftlichen Entwicklung. Eine Untersuchung über Unternehmergewinn, Kapital, Kredit, Zins und den Konjunkturzyklus*, Duncker & Humblot, München/Leipzig.

The Economist 2012, 'A third industrial revolution', in *The Economist – Special Report Manufacturing and Innovation*, vol. 403, no. 8781, pp. 3-6.

SMALL IDEAS, BIG OPPORTUNITIES

FABLAB AT VIGYAN ASHRAM PABAL, INDIA

YOGESH RAMESH KULKARNI

Figure 1: Students carrying a geodesic dome model for low-cost housing (Source: Photography by Vigyan Ashram).

1. Introduction

The FabLab at Vigyan Ashram is located in Pabal, a small drought-prone town near Pune in India. "Pabal is a village of approximately 9,000 people, but because it is a primarily agricultural community, most of the population is scattered outside the core of the village. It is located five hours east of Mumbai by bus; the closest major city is Pune, which is two hours away" (Wagh 2007). In Pabal the Ashram is engaged in training rural youth and school dropouts in various rural technologies and helps them to establish their own enterprises. The FabLab at Vigyan Ashram was established in 2002, with the support of MIT, so it is one of the oldest FabLabs in the world. Furthermore the presence of almost all of the issues related to 'development' in Pabal makes this FabLab a unique one: Inconsistency in supply of electricity, problems of housing, modes of transportation, means of communication, issues of employment, agriculture, sanitation etc. all exist in Pabal at the same time. However the founder of Vigyan Ashram Dr. Shrinath Sheshagiri Kalbag did not only see negative circumstances for establishing a FabLab, but called rural areas goldmines of opportunities for creative scientists, where one can apply simple science and get good results (Kalbag 2003). Kalbag, a scientist turned educationalist, started the Vigyan Ashram in 1983 to find solutions to problems in education. In Sanskrit 'Vigyan' means 'search of truth' and 'Ashram' symbolizes 'a simple living and high thinking' for us this means, that the Ashram is also an organization where all are equal, therefore we call it a 'modern version of the old Gurukul[1] system'.

Vigyan Ashram is a center of Indian Institute Of Education (IIE) Pune. As part of their hands-on training, students provide various technological services to the community. They are required to work on real-life problems in the community as project work. Students are trained to look at the problems as opportunities, and they get experience of thrill in taking up a problem from real life and then inventing a technique to tackle it. An especially designed program for learning how to solve such problems, called Vigyan Ashram's Introduction to Basic Technology (IBT), is accepted by the state government and getting implemented in 122 formal schools, training more than 8000 students every year. In this program, basic hand tools and conventional machines for fabrication, agricultural tools, electrical instruments etc. are provided to schools. Using these basic tools, they work on various projects. Therefore the FabLab at Vigyan Ashram is a logical extension of its program.

In this paper I explain how a real-life problem can become an educational activity for children, leading to research and innovation. Furthermore I present some of the many interesting projects and innovations that are made by these schoolchildren and school dropouts, such as geodesic domes for low-cost housing, 'mechanical-bull tractors', or earth resistivity meters for groundwater prospection.

---

1 | 'Gurukula' is the Sanskrit word for 'House of the Guru' and gurukuls are school types that exist since the Vedic age, where students live in a common place with their 'guru' or 'master'. In a gurukul, the shishyas (students) live together as equals, irrespective of their social standing, learn from a guru and therefore help the guru in his or her day-to-day activities.

2. Small Ideas, Big Opportunities

In our courses the students are encouraged to brainstorm the problems of their everyday lives, at the workplace of their parents, or in the community. Based on these ideas, projects are formulated and technological solutions are developed by students and/or teachers. Within this framework the FabLab is mainly used for automation, electronics and energy-related projects at the ashram. The following examples explain how the work in the FabLab contributes to handling grass-root problems.

2.1 Rural Lighting

In India, the lack of continuous electricity is a major problem in rural areas. Most of the time, electricity is not available at night at all and power outrages during daytime are the normality. Vigyan Ashram got introduced to the use of LEDs in lighting. Engineers in the FabLab developed basic circuits and the design for LED light units. They made different circuits suitable to different specifications viz. 6V, 9V, 230V AC, torch etc.. Furthermore a manual was developed with the list of its parts and the designs of the circuits. Kits for electronic components were made with the help of local electronic component suppliers. Besides also students from schools are encouraged to use the design and make their own LED light unit for their own homes and for their neighbors. At best the students are even able to earn money by selling these self-made units to the villagers. Students came up with many new ideas to reduce the cost of the kits and electronic units. They proclaimed that instead of using the expensive readymade casings, they better used local material like coconut shells, broken pipes, old PC mouses etc. for casing. Last year, students sold more than 5000 units of LED lights.

Problem	Activity	Use of FabLab	Topics covered in the curriculum
Household lighting	Survey needs, the selection of circuits, soldering, testing, installa-tion	• preparing a basic design for a circuit • preparing a manual for LED units • training	• Ohms law, electric circuits • solar energy • calculating energy requirement • design / art / creativity

Table 1: Curriculum for LED units.

As shown in Table 1 the students work on LED lights as part of their school curriculum to learn different curricular areas, such as physics, mechanical engineering and art. Nevertheless, for us the most important aspect of these curricular activities is that the knowledge gained by students does not remain passive but that they are capable to use it to solve their own problems.

2.2 Human Power Energy Devices

Initially, solar and grid power were used to charge LED lights. It was soon found out by the students from tribal and rain-fed areas that solar charging was not possible during four months of the year, during monsoon season. The sun does not shine regularly and strong throughout this time of year and grid supply is also highly unreliable during rainy days. These realizations lead to the initiation of the pedal power generator. The fabrication of the cycle frame was made in a fabrication workshop and the electronic charging circuit was made by using a FabLab machine. The project even won the World Bank's Development Marketplace Competition 2007 award and this solution was supplied to 42 boarding schools for tribal students situated in remote areas.

Ashish Gawade, an entrepreneur from Pune, took the idea further and developed a commercial version of pedal power, forming the company M/S Bottom of the pyramid Energy and Environmental solutions Pvt Ltd in the process.

Through Table 1 and 2 and its explanations the character of the courses and projects of Vigyan Ashram is introduced, but since a detailed description of all of the Ashram's projects would go beyond the scope of this text. Other projects are presented only in the table form.

In contrast to other educational approaches the problems that are solved and the research that is done at Vigyan Ashram does not come as an 'unnatural' assignment from a particular discipline like mathematics, science, etc. and its textbooks, but from the students' reality. Therefore, the students are taught to handle the problem in practical hands-on and interdisciplinary ways. Some of the noticeable projects under development in the FabLab are explained in Table 3.

Figure 2: Human power energy device in form of a bicycle
(Source: Photography by Vigyan Ashram).

Problem	Activity	Use of FabLab	Topics covered in the curriculum
Charging of LED light units during monsoon / rainy days	Fabrication of • a cycle • a charging circuit • pedaling and charging • battery maintenance	• Charging circuits • Preparing a Manual • Testing of cycle and generator	Cycle, levers, gears, power transmission, environment, friction, health, calculating own power

Table 2: Curriculum for human power energy devices.

Name	Problem	Use of FabLab	Possibilities
1) Admin device for teachers	Teachers spend a lot of time keeping records and preparing different reports, whereby the (non) availability of computers is a problem.	To prepare a handheld device to record the students' data. Data can be then transferred to a computer.	One handheld device per teacher.
2) Leaf wetness sensor	Devices available in the market are loaded with unnecessary features, which make them costly.	The circuit produced in the FabLab (which fulfills the needs of grape farmers) has only one essential feature, what reduces the expenses drastically.	Designing a low cost device for the specific requirements of a particular crop. It is useful for disease forecasting.
3) Hearing aid	A villager in Pabal is in the need of a hearing aid.	To produce the electronic circuit and to prototype it.	Develop a DIY design for others to use it.
4) Solar egg incubator	Available hatcheries are usually very big in size and number of units. A further question is how to supply a small-scale hatchery with power, when the use of a diesel engine-generator is not a viable idea	Automation in the incubator to rotate the eggs, maintain the temperature and humidity parameters are build in a FabLab.	The solar egg incubator can work as a livelihood opportunity for women's self-help groups.

Table 3: Overview of FabLab related projects.

3. Difficulties and Solutions

Within the last decade, since the FabLab's establishment in 2002, many interesting projects were realized and (hopefully) many students felt empowered through their experience in the FabLab. But for all that the FabLab story of Vigyan Ashram is also a story of specific Pabal-related obstacles that are probably rather unknown for many other FabLabs. In the next passage I note some of the major difficulties that Vigyan Ashram faced in its operations.

3.1 Software Difficulties

In the initial period of our FabLab, the existing software was not fully developed and it was often advised to again update the software, but downloading updates was frequently a problem due to limitations of the data transfer rate at Pabal. By the time new versions became available, the students came across some difficulties in using it. Now the software has been further developed and has become much more user-friendly – even for newcomers.

3.2 Machinery Problems

Not only did the software cause problems, but also the use of the machinery. Due to voltage fluctuations, laser cutters had problems and motherboards needed replacement twice. The problem was resolved by working with engineers from a laser-cutting machine manufacturing company.

Furthermore, in India, 230V AC supply is used, but some of the machines in the lab run only on 110V supply. Due to a 230V input, some of the tools got damaged. Moreover, different switchboards, color codes etc. are now used to reduce the frequency of such damages. Some of the initial tools were replaced with equivalent Indian-made tools.

Moreover, the CNC plasma cutting table from Torchmate proved to be a problematic machine from the beginning on and we could not really use it. Later, even MIT removed Torchmate from its inventory. Although Torchmate is still present in Vigyan Ashram, we are not able to use it.

In addition to these problems, machines in the lab are most suitable for electronic projects. Most of the requirements from villagers are for agricultural tools, which need metal work and fabrication. Therefore, conventional machines and hand tools still are extensively used in the FabLab. This is also due to the fact that the staff feels more comfortable to use the conventional tools. However, we feel that the use of the machinery will increase eventually.

On top of that, students in Vigyan Ashram are mainly school dropouts. For most of them, their educational language is their mother tongue, which is often not congruent with the 'computer and software language', English. Moreover, most of them saw computers for the first time when they started working at Vigyan Ashram. To introduce them to the machines and to keep the lab operational, we need polyglot electronic engineers as FabLab managers. Finding a suitable FabLab manager, however, has always been a problem.

Lastly, we do not really contribute to an exchange of files and project reports amongst the FabLabs. However, the process of documentation and making design files which can be available for download is in progress. Some of the projects are uploaded on the Internet. Yet, a lot more work needs to be done.

4. Conclusion

It is said that creativity does not depend on availability of tools and infrastructure and many innovations at Vigyan Ashram prove this statement. But with the introduction of the FabLab and tools for digital fabrication, a sense of empowerment, setting aspiration for sophistication, precision and quality are achieved. The glamour of the FabLab machines motivates the youth to become creative and come up with possible things that can be done with these machines. Moreover the ashram has set up a number of regular activities to further serve the community of Pabal: In the last five to six years, almost all new vehicles in Pabal and the surrounding villages are fitted with number plates cut on vinyl cutting machines; All awards, trophies and mementos are regularly cut on laser machines at Vigyan Ashram; Screen-printing and other small projects are the routine introductory activities for new students.

Mahatma Gandhi's idea of self-sufficient villages initiated the use of local resources to meet local needs. Digital fabrication and FabLab networks can make it possible to manufacture most of the goods locally, and the knowledge can be imported or exported. This will create local livelihood opportunities and decentralized production. The concept has the potential to facilitate reverse migration from cities to villages for employment. Kalbag (2003) in this vision states, "my vision is that every village will have the equipment and the skills that are the basis of all modern industry". He expects the youth to become inventors of their own, being able to use a variety of skills and to concretize their ideas even at home. We believe FabLabs have the potential to achieve this vision and therefore Vigyan Ashram is committed to the FabLab movement.

Acknowledgement

Vigyan Ashram thanks Dr. Neil Gershenfeld and the team at Center for Bits and Atoms at MIT, Dept. of Science and Technology Govt. of India for their support.

References

Kalbag, SS 2003, *Selected essays of Dr. S. S. Kalbag on education, technology & rural development*, Vigyan Ashram, Pune.

Wagh, P 2007, *An Exchange of Ideas is Pabal. Defining Development*, viewed 2 January 2013, <http://web.mit.edu/mitir/2007/spring/india.pdf>.

AFFORDABLE AND ADJUSTABLE PROSTHESES CREATED IN FABLABS

ALEX SCHAUB

Figure 1: Prosthesis production (Source: Photography by Alex Cuff).

1. Introduction

The bi-annual International Trade Show for Prosthetics in Leipzig, Germany, shows all the high-tech breakthroughs and innovations in the field of prosthetics at the highest international level. However, for most people in developing countries in need of a prosthesis, these technologies are out of their reach as they are unaffordable. Yet, in one corner of the tradeshow, there are a couple of square meters dedicated to low cost prosthetics. The Red Cross, for example, has an affordable prosthesis, which is even sustainable. One of the questions raised by Waag Society (an Institute for Art, Science and Technology) in Amsterdam, The Netherlands, and The House of Natural Fiber's (HONF) FabLab in Yogyakarta, Indonesia, is, if it is possible to produce a sustainable, high-quality, low-cost below-knee prosthesis, which could be produced in local FabLabs, using local materials. This article describes the efforts of the $50 Prosthesis Project, a collaboration between HONF and Waag Society aiming at creating affordable and self-adjustable prostheses for people in Indonesia.

The House of Natural Fiber's FabLab has initiated a number of projects in the surrounding area, ranging from arts and design to education and public services. In line with its consistent focus on interactivity between people and environments, HONF selects and structures its projects based on the needs of local communities. One of these projects includes research on production and fabrication processes in relation to such fields as robotics, Open Source, and sciences (e.g., microbiology). One of the partner organizations that benefits from the support provided by HONF is Yakkum, a rehabilitation center for disabled people. HONF has been collaborating with Yakkum for almost nine years, working as a non-official mediator and facilitator through workshops in the field of arts and empowerment. The collaboration with Yakkum confronted HONF with its biggest challenge in the context of fabrication processes. Yakkum also produces prosthetics and orthotics for people with physical disabilities, particularly in Yogyakarta and other urban areas in Indonesia. However, these medical aids are expensive to produce, and the making takes much time, seeing as it takes two weeks to finish just one prosthesis. The situation is particularly problematic since there are many patients who urgently need prostheses, and most of them come from poor families. The aim of the $50 Prosthesis Project is to enable Yakkum to provide prostheses for ten people a day using FabLab technology.

2. How to Build Prostheses?

The development of a high-quality, low-cost prosthesis, which is self-adjustable enhances the quality of life of the user, particularly in combination with the user's possibilities of adjusting the prosthesis him or herself. For the design process the knowledge of a user is implemented in an iterative process, and the workplaces of FabLabs are furthermore linked with experts, hence knowledge can be exchanged through the already existing communication channels (a videoconferencing system has been set up among the cooperating FabLabs).

2.1 Challenges in Building Prostheses

The focus of the developing process lies on:

- Lowering the costs of the prosthesis (sustainable, local materials);
- Exchanging knowledge on the process of prosthesis production;
- Applying the user's expertise and involving them in the process of producing affordable health solutions;
- Combining the expertise of both the 'real' professionals and the 'user professionals' in developing an adjustable prosthesis that improves the user's quality of life.

2.2 History of the $50 Prosthesis Project

The first step in this collaborative process took place in May 2009, when the Waag Society's FabLab of Amsterdam invited HONF to an introductory prosthetics workshop for an initial exchange of experiences between users and designers. The workshop covered methods, techniques and materials and included expert input from Hugh Herr, the director of the Biomechatronics Research Group at Massachusetts Institute of Technology (MIT), and Marcel Conradi, the director of the De Hoogstraat Rehabilitation Centre in Utrecht. In addition Appie Rietveld, initiator of Korter maar Krachtig, a Dutch support and advocacy group, provided an end user evaluation for people dealing with limb loss.

A second prosthetics workshop in January 2010 aimed to define design parameters for adjustability, to devise inexpensive, efficient methods for production, and to explore the use of local materials – for instance, using local bamboo instead of aluminum reduces production costs considerably. A very useful insight emerged, when we discovered that the patent of the 'pyramid adapter', a crucial part of the prosthesis, was expired, which allowed the collaborating partners to re-engineer it. The next step was to test a first bamboo prototype and to make it adjustable. Most prosthesis users currently depend on orthopedists for every minor adjustment of their prosthesis, but that could theoretically be avoided. Many users do not realize that they already have a lot of first-hand knowledge about their own prosthesis, since they wear them 24/7; they are the experts at their own prosthetics use. Children generally need to have their prosthetic legs recalibrated by a doctor every six months. In Indonesia, this costs a lot of time and money. An adjustable leg would enable end users to adjust their prosthetic legs themselves by feeling and experiencing the fit, measuring the prosthesis and adapting it.

Walking on different surfaces also requires adaptation of the leg. The roll-off curve of a foot changes drastically when walking on different surfaces. The majority of prostheses on the market are designed for just one standard surface. An adjustable prosthesis would enable users to manage aspects like the roll-off curve, the angle of the foot or the height of the prosthesis themselves. In Indonesia, prosthesis alignment is mainly done manually. To facilitate the process, the collaboration team started to develop tools, such as a cheap alignment laser device and a portable 3D scanner. As DIY kits, these tools could improve accuracy while remaining affordable and accessible. Besides using digital fabrication resources, the team embraced open innovation principles, like pooling knowledge from the expert users in Yakkum, the designers from HONF and FabLab Amsterdam, academic advisors such as Prof. Dr. Bert Otten (Center for Human Movement Sciences, NeuroMechanics and Prosthetics, University of Groningen) and specialized manufacturers like Orthopedietechniek De Hoogstraat in Utrecht. Input from all the parties was used in the process of developing and designing the adjustable leg.

The concrete results of the $50 Prosthesis Project so far also include key design insights: For instance, adjustability allows end users to take a crucial step towards independence, and the visual design of the prosthesis is important to end users. In addition, knowledge transfer during production is important for empowerment and self-reliance. In terms of production, the team gathered knowledge on how to use thermoforming to produce quality limb sockets quickly.

2.3 Future of the $50 Prosthesis Project

The next steps address specific, tangible end user needs and preferences. What do users need in order to adjust the prosthesis effectively? How would they like the design to look and feel? The aim is to develop a process or method for design based on the parameters defined in consultation with 'expert users': adjustability, open innovation and digital fabrication. To this end, a FabLab is set up in Yogyakarta with a special prosthetics section. However, it is a challenge to set up such an entity in a developing country and to make sure it is sustainable. After the pilot phase is over, the aim is to have a business plan to ensure that the FabLab could work as an independent digital fabrication workshop. The work on the $50 Prosthesis Project will hopefully continue after the end of the pilot phase.

The ambitions are to have a FabLab hub on prostheses, where research and practice are united. An online community was created (www.lowcostprosthesis.org), where experiences can be exchanged and case studies can be gathered so as to minimize the chance of reinventing the wheel. In the future, we plan to research options for using intelligent materials to enhance the experience and effectiveness for the end user. Another goal is to explore the use of embodied cognition. Embodied cognition means that aspects of the body shape aspects of cognition. Bert Otten expects the process of prosthetic design to be guided by the team's increased insight into the development of embodied cognition of physically challenged people as they learn to walk with the leg prosthesis. The physically challenged people's improved sense of dynamic balance can be observed best from the way they move and how they intuitively adjust their prosthesis. Consequently, no technical insight or expertise should be needed to adjust a prosthesis optimally, as long as the design is based on embodied cognition. This way of implicit learning is crucial. Learning to walk with a prosthesis is a sensitive process. Showing the user a video of his or her best performance is more effective than giving orthopedic advices or showing figures.

Recently, in the summer of 2012, we have been experimenting with composites containing natural fibers and Polyurethane plastics. With very fine pineapple tree fibers combined with a PU rubber compound, a surprisingly strong stomp socket could be completed (see Figures 2 and 3). The resistance and durability has yet to be thoroughly tested. In Indonesia crafters are prototyping sockets using the

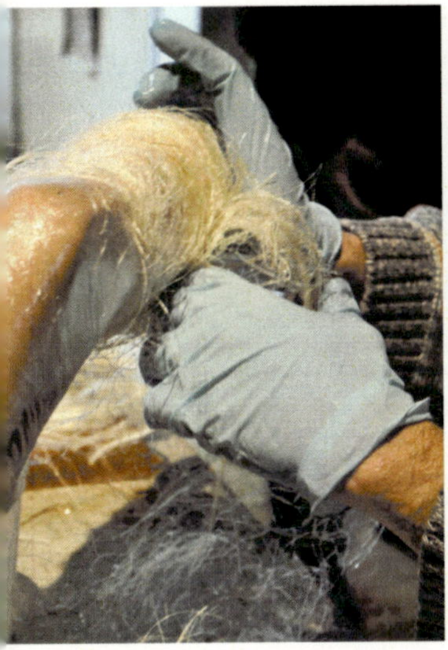

Figure 2: Pineapple socket (Source: Photography by Mickael Boulay).
Figure 3: Vessels (Source: Photography by Alex Schaub).

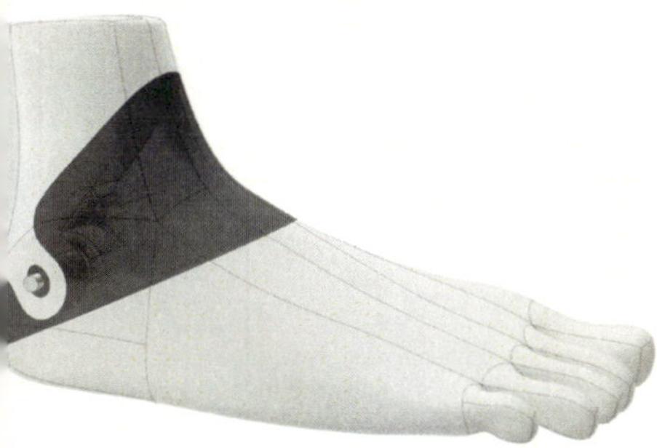

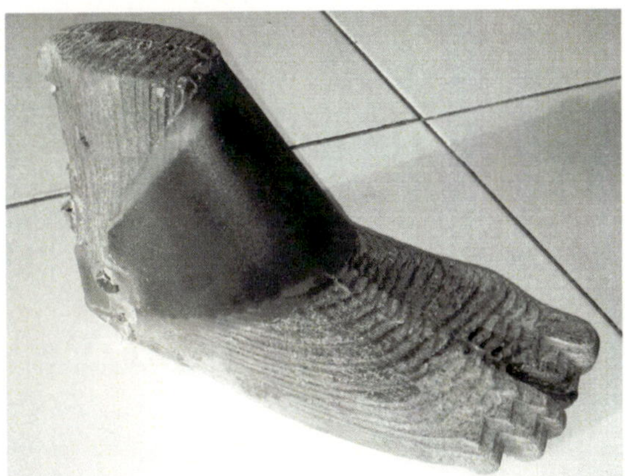

Figure 4: Wedge foot sketch (Source: Illustration by Jens Dyvik)
Figure 5: Foot prototype (Source: Photography by Alex Schaub).

traditional craft of bamboo coiling. Designer Mickael Boulay is currently prototyping an adapter that intuitively allows the user to adjust his or her feet and knee angle to his or her needs.

Another prototype we are working on is a new foot design. 3D milled wooden parts are combined with Polyurethane rubber to make a flexible, yet adjustable foot (see Figures 4 & 5). A second foot design approach is based on a molding and casting technique. Negative molds are casted on 3D milled wooden foot positive molds. Then the negative molds could probably be stuffed with a mixture of natural fiber and ideally natural rubber.

Further investigation goes into the bio data of the natural fiber itself. One of the most explosive questions we are researching at the moment is called, is it possible to find some oils within the biological structure that can be used as a binder for new resin types?

3. Conclusion

This article described why there is a need of self-adjustable, low-cost prostheses. And how Waag Society's FabLab Amsterdam and The House of Natural Fiber's FabLab Yogyakarta collaborate using open innovation system. Furthermore, it explained the challenges and planned future developments for improving the prostheses (development). All of the described prostheses' designs are based on locally available materials for low or almost no cost. The goal is to increase physically challenged peoples' awareness of their own prostheses, what will improve their daily experiences. Controlling their own prostheses will not be dependent on expert knowledge, but will be based on the physically challenged people's experiences and their empowerment. We believe this will have a big impact on society – everywhere.

EPILOGUE

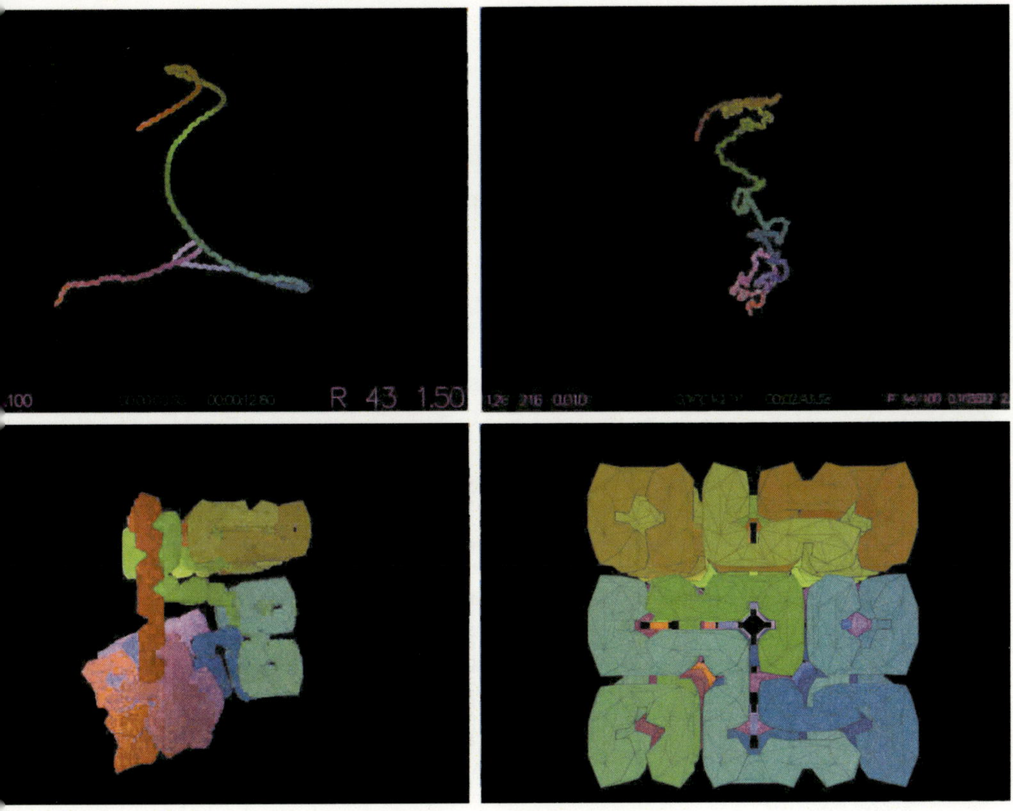

Figure 1: Computational model of digital materials.
A string of discrete elements, each element containing communication and actuation capability, is coded to self-construct into a cube and deconstruct again to its native state
(Source: Model by Jonathan Bachrach).

FABLABS: THOUGHTS AND REMEMBRANCES

SHERRY LASSITER

1. Introduction

A sheep construction kit... an alarm clock that steals your sheet in order to get you out of bed... an elegant electric guitar... a 3D printer... a personalized solar house optimized for your longitude and latitude. What do these whimsical, practical and futuristic things have in common? They were all made in a FabLab. A FabLab is a fabrication laboratory, or as many of us like to say, a FABulous laboratory – a giant playroom on digital hormones, hosting a sophisticated suite of computers, electronics, tools and industrial manufacturing machines. It is a prototyping platform sporting a high resolution CNC milling machine, a laser cutter, a sign maker, a 3D printer, and a large scale CNC router table, accompanied by nano-second time resolution electronic components and programming tools, supported by Open Source and free software and tools, including some written by researchers at the Massachusetts Institute of Technology's (MIT) Center for Bits and Atoms (CBA). This conglomeration of digital fabrication tools and processes allows anyone, almost anywhere, to make almost anything. It is an inspirational environment, for education, for innovation – a creative arena for personal expression.

CBA is an interdisciplinary research program at the MIT established in 2001 by a large National Science Foundation grant to explore the merging of physical and computer sciences. CBA frames its research around the idea of digital revolutions. We posit that the first stage of the digital revolution revolved around digital communications, based in great part on the work of Claude Shannon. Shannon's information theory, as described in his publication A Mathematical Theory of Communication, defined bits, as well as laying the foundations for data compression and error correction in communications (Guizzo 2003). Thus, in part due to Shannon's error-correcting signals, in the 1940s we progressed from analog to digital communications. The second stage of the digital revolution started in the 1950s around computation, based on some of the work pioneered by Vannevar Bush a few decades earlier. Among a number of inventions, Bush created a differential analyzer – a mechanical analog computer that could solve differential equations up to the sixth order (Tamir 2008). It was an enormous device, filling an entire room, and the joke at MIT is that the longer the machine ran, the worse the answers became.

We are in the third stage of the digital revolution now, emerging from analog to digital fabrication – moving from a world populated by large factories full of equipment, often robotic, to digital materials; that is, elemental materials that one can program or instruct to assemble themselves into different forms and functions. An added advantage and promise of digital materials is that you can instruct the materials to recycle themselves into another form and function once an object outlives its usefulness. This is our focus at CBA, the research space in which physical and computer science merge, where the digital bits of the world meet the physical atoms and form and function become one and the same. We are borrowing ideas from biology to guide us. Very simply, a human starts life as a lively mix of basic building blocks, like amino acids and proteins. If you send codes of instructions, RNA and DNA, into this rich mix of fundamental elements, the amino acids assemble into chains as proteins, into cells, organs – eventually growing into a human form and function. Similarly, we are creating fundamental digital materials, self-assembling building blocks, sending programs for self-construction into the materials to create specific forms and functions. As we push the boundaries of this physical frontier, we also explore new computational methods and architectures. And of course we make things, lots of things, from the nano-scale all the way up to the building scale (see Figure 1).

The CBA was funded in 2001 by the National Science Foundation to conduct research at this digital-physical boundary. The funding helped us complete work already in process to build a state-of-the-art digital fabrication facility for research. At the time, the facility included CO_2, femtosecond and excimer laser systems, an environmental scanning electron microscope, a focused ion beam, a waterjet cutter, 3D printers, high-resolution NC milling machines, NC knives, large scale routers, CNC lathes and much, much more. We were in a position to make, literally, almost anything. This facility was somewhat unusual in that it housed all of this capability under one roof, rather than spreading it across MIT's campus based on siloed disciplines. The proximity of all these tools and capabilities inspired MIT students – they wanted access to these powerful tools, not solely for research, but for personal expression. Thus the now famous How To Make (almost) Anything (HTMAA) class was born. Over the course of one semester HTMAA, students would learn all the tools and processes in the facility and as final assignment, make a functional system integrating what they had learned. Students have been wildly creative, and over the years this class has produced everything from ScreamBody, a device that records and saves screams for later release in a more appropriate environment, to a Web-browser for lonely parrots to visit other lonely parrots on the Internet, to a dress that protects your personal space by deploying spines to keep unwanted suitors at a safe distance, to a DIY routing machine made from Lego constructed circuits, and very aggressive alarm clocks for sleep-deprived students. And as students made things that were personally meaningful, they pulled for advanced technical knowledge as they needed it and applied that knowledge immediately. Also known as 'just-in-time' learning, HTMAA created a very powerful learning experience for all.

The level of national public funding that opened CBA's doors required community and educational outreach on our part. At CBA, we understood making things and decided that our outreach would be to take a subset of the digital fabrication equipment, tools and processes used in our research facility and put it into communities to see what people would do if given access to the capabilities and technologies to make almost anything. These FabLabs were designed to be prototyping platforms for grassroots invention, but are now increasingly being adopted by schools and community centers for education and for science, technology, engineering and mathematics (STEM) outreach. There are a growing number of Senators and Representatives in the US Congress who envision a national network of FabLabs as stimulus for innovation and as platform for STEM education and workforce development, an important foundational investment in technical capacity-building that could help ensure US global competitiveness.

2. Applications and Implications

Figure 2: FabFi uses common building materials and off-the-shelf electronics to transmit wireless signals across distances of up to several miles in Jalalabad, Afghanistan (Source: Photography by Amy Sun).

The concept of a distributed network of digital fabrication laboratories has long held the promise of collaborative research and innovation, a global knowledge-sharing, knowledge-building ecosystem. The FabLab network is growing into that potential, with a number of collaborative research projects and innovations rising to the surface.

FabFi is a worldwide FabLab collaboration to bring Internet access to anyone, anywhere for both content creation and content consumption. FabLab partners in Greece developed the first antenna design; in Norway the design was given its form factor. South Africa fabbers tested point-to-point communication capabilities. Indian fabbers worked on distance capabilities. The parabolic antenna design for the repeaters was developed at MIT. And corporate partners in Boston adapted routers and networking protocols to support the system. The first fully functional FabFi network was deployed in Jalalabad, Afghanistan in 2009 by a group of local FabLab users, alongside an international cohort of FabLab volunteers with engineering, IT and software development expertise. FabFi is a project of Fab Folk Inc., a US-based non-profit organization that supports FabLab users around the world, run by MIT Ph.D. graduate Amy Sun. FabFi is now working with Kenyan FabLab partners to develop Open Source business models that will cover billing and technical network management. All of the designs and management information are freely available through Open Source or creative commons licensing. Though less active now than in earlier years, it is a beautiful example of global collaboration, development and innovation.

The Fab Lab House (http://www.fablabhouse.com) was another global collaboration led by the Institute for Advanced Architecture of Catalonia (IAAC). Originally created as an entry to the European Solar Decathlon Competition, the design, interior and exterior of this solar house were made in the Barcelona FabLab, with the help and collaboration of FabLabbers from Spain, Australia, Lebanon, Mexico, Peru, Ethiopia, USA and more. It was designed to be customizable and energy self-sufficient. The parametric modeling tool optimizes the FabLab House design for light capture based on the longitude and latitude of the building site. Earning the People's Choice Award in the competition, the team plans to franchise the model for design and construction across the globe. This approach allows for rapid prototyping of personalized housing locally and has implications for relief efforts and housing shortages across the globe.

The cost of buying and shipping a FabLab for many communities is prohibitive, on the order of $100,000 for the full-scale model. Many individuals at MIT and within the larger FabLab community are working to reduce the cost and complexity of getting a FabLab. The Machines That Make project (MTM) at MIT was created partially as a response to this need, building on a worldwide effort to bring the tools for digital fabrication within reach of anyone, anywhere. The MTM Snap is a high-resolution milling machine that can be made entirely in a FabLab for one tenth of the cost of the industrial model it replaces. The MakerBot and Ultimaker DIY 3D printers can also be made in a FabLab, again for about a tenth of the cost of their industrial counterparts. Of the entire FabLab toolset, we anticipate that within three to five years, a full-scale FabLab will be capable of completely reproducing itself; that is, the machines in a FabLab will be used to make entirely new FabLabs at a tenth of the cost. Rather than exporting expensive equipment using fossil fuels for transport and incurring heavy import duties with foreign customs agencies, we can ship the files to make a FabLab over the Internet for free and the FabLab can be manufactured locally for a tenth of the cost. This is a fun twist on the sci-fi horror vision of self-replicating machines, but there is no doubt that friendlier self-replicating FabLabs are well within reach.

The cost and complexity of digital fabrication are coming down very quickly, bringing the tools and capabilities for invention and innovation within the reach of almost anyone, almost anywhere in the world. That kind of democratic, distributed capability is revolutionary. It promises to change everything... the way we design, manufacture, finance, communicate and market for business, supply chains, the platforms upon which we build businesses, the way we educate and conduct research. And a pioneering group of fab individuals is constructing foundations to leverage and support the digital revolution in fabrication, foundations that merge design, Open Source collaboration, social networking, entrepreneurship and digital fabrication into something new and electric. In the preceding pages, you can sample some of these special individuals and the contributions they are making to our global future.

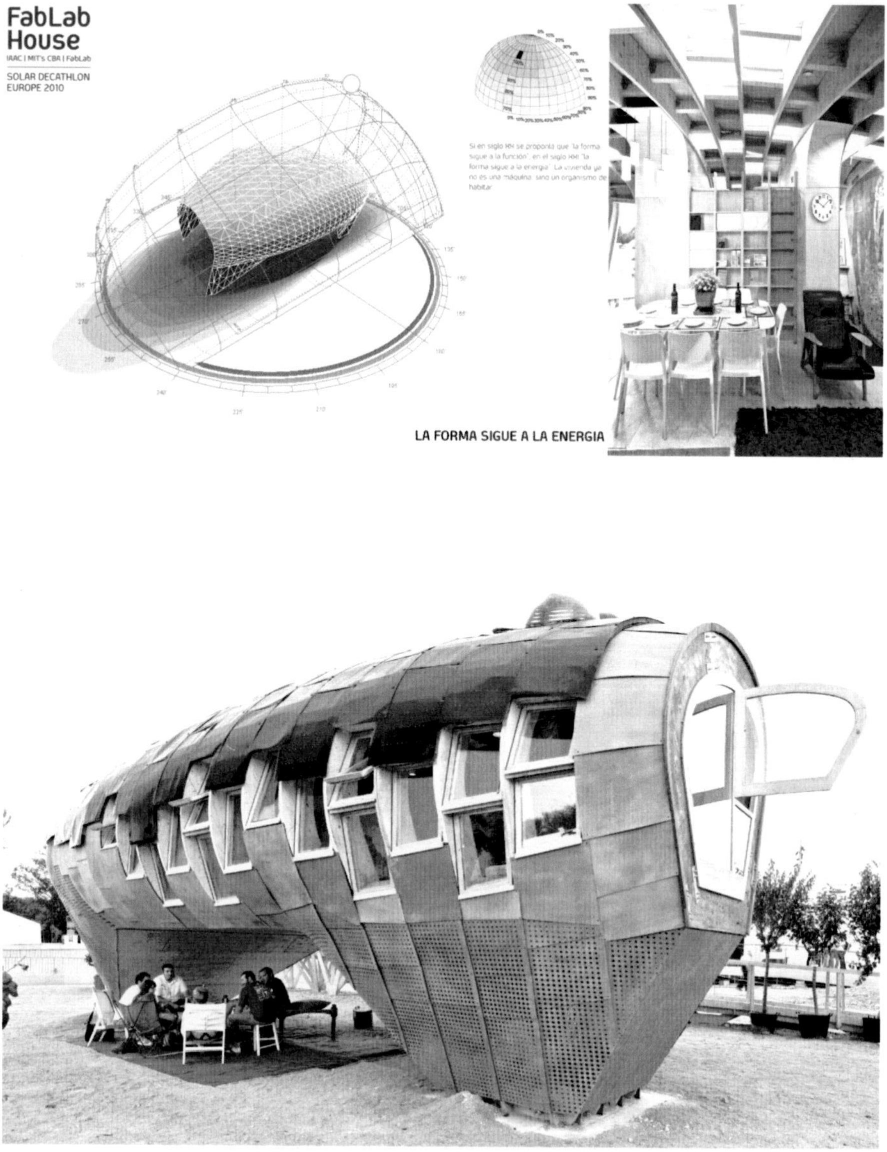

Figure 3: Model of the Fab Lab House and inside and outside perspectives of the Fab Lab House (Source: Design and photography by Adrià Goula).

3. Education

The FabLab by both design and serendipity is a trans-disciplinary, catalytic environment for applied learning in STEM disciplines. Many of us in education hear the overused terms 'hands-on' and 'project-based' learning, but FabLabs are where you see the magic of applied learning in action. The excitement and empowerment of making a functional system all by yourself (with a little help from your friends) cannot be overstated. FabLabs employ Seymour Papert's constructionist approach to teaching and learning. Constructionism uses Piaget's theories of constructivist learning as a foundation – learners actively construct knowledge from their experiences in the world – but builds on that foundation by recognizing that new knowledge is constructed more effectively when the learner is engaged in making things that are personally meaningful (Resnick 1994, pp. 23-24). Although implemented primarily in community development and informal education contexts, recent FabLab installations in community colleges and in a handful of public schools have extended the constructionist approach into formal education institutions, thereby enabling a pioneering community of children and young adults to build new literacies that help them shape their digital and physical worlds. There are networks of schools in the USA reaching out for FabLabs as STEM learning becomes part of the national agenda. Abroad there are a handful of adventurous schools that are adopting it as well, and the requests for more FabLabs in educational settings are mounting into the hundreds.

Formal education institutions aside, there are innovative educational approaches emerging from the network that have interesting implications for the future, notably the Fab Academy and the Green Fab Project. Led by MIT Professor Neil Gershenfeld and Vicente Guallart of the Institute for Advanced Architecture of Catalonia (IAAC), Fab Academy utilizes the distributed nature of the FabLab network to offer affordable advanced technical education to a global campus. Peers in FabLabs work together with local technical mentors (or gurus as we fondly call them) connected by Internet-based videoconferencing, communications, publishing and archiving tools in a global version of the How To Make (almost) Anything class, emerging with interdisciplinary 21st century skills recognized worldwide, and becoming members of the community able to build and manage FabLabs as they scale worldwide.

The IAAC team in Barcelona leads another education innovation called Valldaura Self-Sufficient Labs. The Valldaura project is host to three labs: Food Lab, Energy Lab and the Green Fab Lab. It is dedicated to the education, research and dissemination of innovations in renewable energies and eco-efficiency as they relate to human habitation and architecture. IAAC has renovated an old monastery situated in a forest on the outskirts of Barcelona to serve as a platform for the development of research projects on multiple scales (city, building, manufacturing) and in different areas of expertise (ecology, energy, digital manufacturing, and new technologies). Valldaura Self-Sufficient Labs will merge ancient practices in agriculture and fabrication with state-of-the-art digital fabrication technologies in the search for efficient, sustainable solutions for our future world. The FabLab network will disseminate and scale the most successful innovations as well as the innovation model itself.

Figure 4: Valldaura Self-Sufficient Lab and FabLab-made solar cells (Source: Photography by Sherry Lassiter & Susan Klimczak).

CBA, through National Science Foundation funding, built the first ten to fifteen FabLabs in the world, starting in rural India, then Northern Norway, then metropolitan and rural Africa, then in the US. There are approximately 150 labs now, all funded and built by communities themselves. The FabLab network is experiencing its own version of Moore's Law, doubling in size approximately every 18 months, so we expect approximately 150 new FabLabs to come online in the next year and a half. This is a distributed network for education, invention, business generation, manufacturing, and research. The articles you find herein tell the remarkable story of the people, the research, and the wildly creative ideas coming from this incredible community.

Please allow me here to acknowledge some of the founders and architects of the FabLab movement. First, Professor Neil Gershenfeld, Director of the Center for Bits and Atoms at MIT, is the visionary who pulled all these great people and ideas together, saw the future, and made it happen. He pushed, prodded, supported, and mentored the FabLab movement into being. Without him, this very special community would not exist. Saul Griffith and Bakhtiar Mikhak are both remarkable researchers who did some of foundational research that kicked off this wild movement. Manu Prakash, Amy Sun, Amon Millner, Kenny Cheung and Nadya Peek are the MIT Fab gurus who literally built the physical and social foundations of the FabLab network. Haakon Karlsen of Norway had a global vision for our knowledge-sharing community and put the humanity into Fab. Hanna van Essen and Klass Hernamdt of the Netherlands understood the value of education and innovation and scaled the FabLab idea across Europe. Vicente Guallart and Tomas Diez from Barcelona are inventing the future with FabLabs, constructing their visionary aircraft as they fly it. And there are many, many individuals throughout the world who have contributed their time, thoughts, labor and passion to the FabLab movement; to them we are deeply grateful.

In closing I'd like to impart the sentiment of my favorite Next-Gen fabber from Boston, Makeda Stephenson to you: "Go forth and make great stuff!"

References

Guizzo, EM 2003, 'The essential message: Claude Shannon and the making of Information Theory', Master's thesis, Massachusetts Institute of Technology, Cambridge, MA.

Resnick, M 1994, *Turtles, termites, and traffic jams: explorations in massively parallel microworlds,* The MIT Press, Cambridge, MA.

Tamir, B 2008, *The differential analyzer,* The future of things, viewed 12 January 2013, <http://thefutureofthings.com/column/1011/the-differential-analyzer.html>.

LIST OF AUTHORS

Bakker, Bart is a technology forecaster from the Netherlands. He is deeply involved with Protospace, Utrecht's main FabLab, and with the Dutch FabLab Society.

Blikstein, Paulo, Prof. Dr. is an assistant professor at Stanford University, USA. He is affiliated to the Stanford University's Graduate School of Education and (by courtesy) the Computer Science Department. He designs and researches expressive technologies for learning, especially for underprivileged populations and he is also involved in the fields of computational modeling and complexity science.

Bohne, René is a research assistant for the Media Computing Group at the German university Rheinisch Technische Hochschule Aachen. He is also the university's local FabLab manager.

Büching, Corinne is a sociologist in the field of science and technology. Her main research interests are qualitative methods. She is a research assistant in the working group Digital Media in Education (dimeb) at the University of Bremen, Germany.

Carstensen, Tanja, Dr. is a sociologist working as a post-doc researcher in the working group Work – Gender – Technology at the Hamburg University of Technology, Germany.

Dittert, Nadine is a research assistant in the working group Digital Media in Education (dimeb) at the University of Bremen, Germany. She is a computer scientist who is also interested in developing learning and teaching scenarios for FabLabs.

Döring, Tanja is a computer scientist working as a research assistant at the University of Bremen at the Center for Computing and Communication Technologies. She is involved in the FabLab St. Pauli, Germany.

Dyvik, Jens from Norway is a cabinetmaker, designer, and blogger, who is currently on a FabLab world tour.

Grosskopf, Lambert, LL.M.Eur., Prof. Dr. is a legal academic at the University of Bremen, Germany and an Attorney at Law, a Certified Lawyer for Information Technology Law (IT-Law) and a Certified Lawyer for Copyright and Media Law.

Joost, Karsten from Bremen, Germany, is an artist, toolmaker and FabLab networker.

Katterfeldt, Eva-Sophie is a computer scientist and a research assistant in the working group Digital Media in Education (dimeb) at the University of Bremen, Germany. Her research interests focus on aspects of modeling in Computer Science, Interaction Design and maker practices.

Krannich, Dennis, Dr. is a senior researcher at the University of Bremen in the working group Digital Media in Education (dimeb). He is a computer scientist who is also interested in setting up a FabLab at his local university.

Kulkarni, Yogesh Ramesh, Dr. is the manager of the FabLab at Vigyan Ashram in Pabal, India.

Lassiter, Sherry is the Director of the international Fab Foundation and Program Manager for the Center of Bits and Atoms at the Massachusetts Institute of Technology, USA. She supports and coordinates the worldwide FabLab network.

Lund, Michael has a background in Art History, Cultural Studies, Media, and Cultural Mediation. He is a research assistant in the working group Digital Media in Education (dimeb) at the University of Bremen, Germany.

Nake, Frieder, Prof. Dr. is a professor emeritus at the University of the Arts, Bremen and the University of Bremen, Germany. His research interests include digital media and digital arts. He is a pioneer in Computer Art.

Pettis, Bre from Brooklyn, USA, is one of the founders and the CEO of MakerBot®, as well as one of the founders of Thingiverse.

Posch, Irene is a researcher with a background in Computer Science and Media. She designs the youth programs at the Happylab in Vienna, Austria.

Robben, Bernard, Dr. is a senior researcher in the working group Digital Media in Education (dimeb) at the University of Bremen, Germany. He has published various articles on media theory, the computer as a medium, tangible embedded and embodied interaction and the design of 'be-greifbar' (tangible and graspable) media.

Schaub, Alex is the FabLab manager of Waag Society's FabLab in Amsterdam and the head of the $50 Leg Prosthesis Project in Yogyakarta, Indonesia.

Schelhowe, Heidi, Prof. Dr. is a university professor for Digital Media in Education (dimeb) and the head of a working group of the same name at the University of Bremen, Germany. Additionally, she is the Vice Rector for Teaching at the University of Bremen.

Sterling, Bruce is a science fiction author and technology essayist from the USA.

Sylvester, Axel is a selfemployed IT consultant and independent researcher. He has a Master's degree in information systems from the University of Hamburg. Sylvester is the manager and one of the founders of the FabLab Fabulous St. Pauli, Germany.

Troxler, Peter, Dr. is a Research Professor at Rotterdam University of Applied Sciences on the topic of the 3rd Industrial Revolution, a freelance researcher at the joint of business administration, society, and technology. He is involved in many local, regional and global FabLab initiatives.

Walter-Herrmann, Julia is a media- and social scientist, working as a research assistant in the working group Digital Media in Education (dimeb) at the University of Bremen, Germany.

Zeising, Anja, Dr. is a research assistant in the working group Digital Media in Education (dimeb) at the University of Bremen, Germany. Her research foci revolve around designing digital media and interactions beyond the desktop with and for children.